# Study Guide

**Michael J. Leboffe**
*San Diego City College*

---

# Biology
## Fundamentals

**Gil Brum**
*California State Polytechnic University, Pomona*

**Larry McKane**
*California State Polytechnic University, Pomona*

**Gerry Karp**
*Formerly of the University of Florida, Gainesville*



John Wiley & Sons, Inc.
New York / Chichester / Brisbane / Toronto / Singapore

ISBN 0-471-07623-6

Printed in the United States of America

10 9 8 7 6 5 4 3 2 1

# Preface

## Study Guide Organization

Although many students anticipate that their first biology class will cover material with which they are already partially familiar, they often find biology to be quite challenging and foreign. There are many new terms and concepts to master, and it's often difficult to know where to begin studying. This Study Guide contains questions that direct your review of important definitions and concepts covered in your text. It also has exercises that give you critical thinking practice by finding relationships between concepts.

Study Guide chapters correspond to chapters in *Biology: Fundamentals,* by Brum, McKane and Karp. Each includes the following components.

- Go Figure! - These are questions designed to clarify and expand upon principles presented in selected text figures. The intent is to focus attention on the value of these pedagogical aids. In some cases the answer may be determined by examining the figure closely. In others, the answer requires interpretation of the figure or application of what you have learned elsewhere.

- Matching Questions - These test your recall of definitions. Most items are highlighted terms from each chapter. Use these questions as a review after reading the chapter and attending lecture. Answers to odd-numbered items are in the Appendix; answers to all questions are in the *Instructor's Manual* available to your professor.

- Multiple Choice Questions - These test your recall of facts and concepts covered in the chapter and also should be used as a review after reading and lecture. Questions are arranged in pairs testing comparable material. The answer is given to the first (odd-numbered) question of the pair. Occasionally, sequences of two or more questions are paired. These sequences are clearly marked and answers are provided to all questions of the first sequence. Answers to all questions are in the *Instructor's Manual* available to your professor.

- Concept Map Construction Problems - These allow you to develop your concept mapping skills. They help you recognize relationships between concepts in the current chapter and between concepts in the current chapter and previous chapters. Because they are comprehensive, attempt these questions as a final review of the material before moving on to the next chapter. Your professor has been provided with a sample concept map for each question in the *Instructor's Manual.* Concept map construction is explained on the next page.

- Answers to Odd Numbered Problems - These include matching and multiple choice questions only. Answers to all matching and multiple choice questions, and sample solutions to concept map problems have been provided for your instructor in the *Instructor's Manual.*

I hope this Study Guide helps you in your biology class. Good luck!

## Concept Map Construction

Suppose you were asked to describe how human, gasoline, motor boat, truck, and vehicle are related. There are lots of ways you could set about doing this. For example, you could write a narrative description, give an oral presentation, or draw each object to illustrate similarities and differences. Another familiar method is an outline. One possible outline is shown below.

- I. Vehicles
  - A. Water Vehicles
    - 1. Sail Boat
    - 2. Motor Boat
  - B. Land Vehicles
    - 1. Examples
      - a. Truck
      - b. Bicycle
    - 2. Wheels
- II. Power Source
  - A. Human
  - B. Engine
    - 1. Electric
    - 2. Gasoline
  - C. Wind

An outline has advantages over the other methods. In a short time, someone can examine it and understand the hierarchical relationships visually reinforced by indentations. But notice many important connections are not indicated. For instance, trucks and bicycles both have wheels, and trucks and motor boats both have engines. Neither is indicated by the outline. Further, it is unclear what power source is used by each vehicle. Because of these shortcomings, other methods have been devised to show relationships which also require a short period of time to understand and visually reinforce associations. One of these is the concept map.

Concept maps are a useful tool for showing relationships between items that might be overlooked or too cumbersome to show in an outline. The idea of a concept map is simple. There are four parts:

1) boxed terms and/or concepts,
2) connecting lines,
3) "connector" words or phrases,
4) cross links.

A hierarchy is formed by placing the broadest (most inclusive) term/concept at the top of the map and more specific terms/concepts towards the bottom. Lines are then used to indicate a relationship between boxed terms/concepts. A brief connector phrase is written over the line to let the reader know what that relationship is. If the relationship is between items in different branches, an arrow is used to indicate the direction of the relationship. Because of their flexibility in showing connections and their hierarchical arrangement, they reflect how information probably is stored in our memory.

Linking terms/concepts is often the most difficult part for beginning concept mappers. Connectors should be nontechnical and usually indicate physical (under, near, connected to), temporal (precedes, follows), or logical (causes, produces) relationships, or subcategories (includes, demonstrates, such as). Keep them simple and do not include words or phrases which could be incorporated as boxed terms/concepts.

On the next page is a sample of a concept map developed from the following terms: human, gasoline, motor boat, truck, and vehicle.

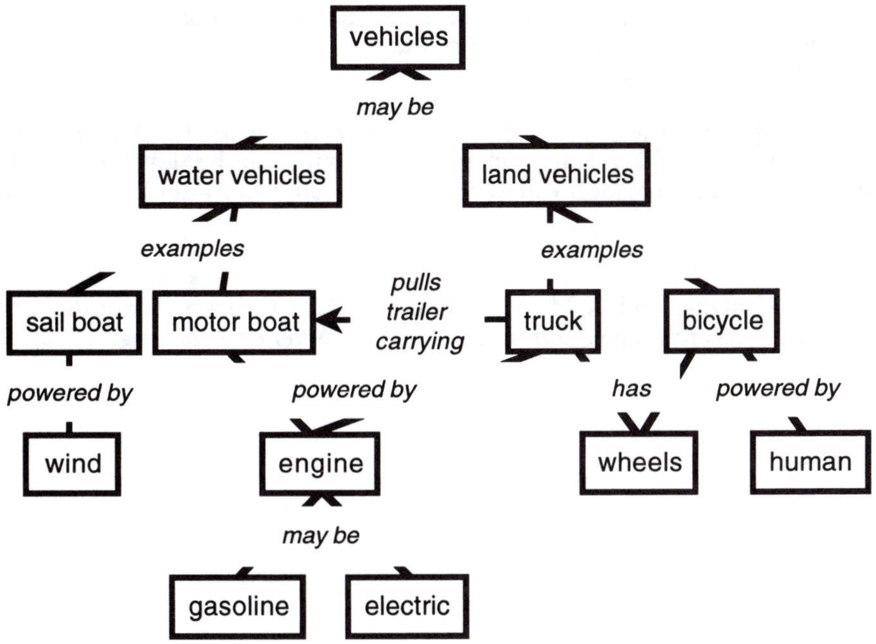


Points of Interest

1. Notice that the term "vehicle" was assigned, but the plural "vehicles" is used in the concept map. Terms will be assigned in the singular form, but it is acceptable to substitute with a plural form.
2. Start at the top of the map and read down any branch. Notice that each pair of terms joined by a line and a connector phrase makes sense. Each pair must stand alone as a sensible statement.
3. Additional terms (water vehicles, land vehicles, bicycle, wind, sail boat, engine, and wheels) were used to clarify and expand relationships. You are encouraged to do this, and it may even be essential to include these intermediate terms to "bridge the gap" between concepts that are not closely related. *Rest assured, however, that any term can be incorporated into any concept map with the appropriate connectors and intermediate concepts.*
4. Notice that lines extending from an item using the same connector phrase arise from the same point on the rectangle (*e.g.,* "water vehicle" connected with "sail boat" and "motor boat"). Lines extending from an item using different connector phrases arise from different points on the rectangle (*e.g.,* "bicycle" connected with "wheels" and "human"). This makes reading the map easier.
5. Notice the cross link indicated by the arrow between "truck" and "motor boat." The relationship only works in one direction (*i.e.,* trucks pull motor boats, not the other way around), so the arrow is essential.

Don't be concerned about constructing the "correct" map; many are possible for any given set of terms. Since there is no single correct answer, other criteria must be used for determining the quality of your map. Often the logic of the hierarchical arrangement, number and quality of added terms, use of cross links, and validity of connector phrases are criteria for evaluation. Mechanical aspects may also be used. These include use of rectangles around concepts, connector phrases, and arrows for cross links. Your professor will let you know how and if yours will be evaluated.

Wallace, Mintzes and Markham (1992)* suggest several "do's" and "don'ts" of concept mapping that our students have found useful. These are:

1. Don't become discouraged; do set it aside prior to the frustration point.
2. Do work with other students.
3. Do seek assistance from your professor or teaching assistant.
4. Do construct concept maps for other courses.
5. Do practice, practice, practice.

Completed concept maps give the illusion they were easy to construct. Our experience has been that novice mappers struggle, but with practice, map making skills can be developed and improved. Once developed, concept mapping provides you with a very useful study tool for any discipline, not just biology.

*Wallace, J.D., J.J. Mintzes, and K.M. Markham. 1992. Concept Mapping in College Science Teaching - What the Research Says. *Journal of College Science Teaching* 22(2):84-86.

# Acknowledgements

This *Study Guide* includes many contributions by my long-time friend and colleague, Gary Wisehart. It seems strange to be soloing after all these years of collaboration, but his effect on this *Study Guide* is still noticeable to those familiar with his style. Thanks, Gary.

Recognition of another long-time friend and colleague, Dr. David Singer, is also deserved for his indirect influence on my development as a teacher. Dave has been preaching emphasis of process over product since we were in graduate school many years ago (long before it was fashionable to do so). Dave introduced our faculty at San Diego City College to concept mapping and the idea for the Go Figure! exercises. This project is better because of his influence.

Thanks and love, too, to my wife Karen for her support and proof-reading, and to my $F_1$'s Nathan, Alicia, and Eric for understanding that "Daddy has work to do." Well, guys, I'm done now...let's go kick around a soccer ball!

# Table of Contents


## Part 1 - Biology: The Study of Life

CHAPTER 1 - Biology: The Study of Life ....... 1

## Part 2 - Chemical and Cellular Foundations of Life

CHAPTER 2 - The Chemical Basis of Life ....... 9
CHAPTER 3 - Cell Structure and Function ....... 17
CHAPTER 4 - The Metabolism of Life ....... 23
CHAPTER 5 - Energy and Life: Photosynthesis ....... 29
CHAPTER 6 - Energy and Life: Respiration and Fermentation ....... 35
CHAPTER 7 - Perpetuating Life: Cell Division ....... 41

## Part 3 - The Genetic Basis of Life

CHAPTER 8 - Fundamentals of Inheritance ....... 49
CHAPTER 9 - The Molecular Basis of Genetics ....... 59
CHAPTER 10 - Gene Expression: Orchestrating Life ....... 65
CHAPTER 11 - Genetics and Human Life ....... 71

## Part 4 - Form and Function of Plant and Animal Life

CHAPTER 12 - Plant Life: Form and Function ....... 79
CHAPTER 13 - Plant Life: Reproduction and Development ....... 87
CHAPTER 14 - Animal Form and Function ....... 95
CHAPTER 15 - Regulating Bodily Functions: The Nervous and Endocrine Systems ....... 101
CHAPTER 16 - Sensory Perception ....... 109
CHAPTER 17 - Protection, Support, and Movement ....... 115
CHAPTER 18 - Nutrition and Digestion ....... 121
CHAPTER 19 - Circulation, Excretion, and Respiration ....... 127
CHAPTER 20 - Internal Defense: The Immune System ....... 137
CHAPTER 21 - Reproduction and Development ....... 143

## Part 5 - Evolution and Diversity

CHAPTER 22 - Mechanisms of Evolution ....... 151
CHAPTER 23 - Evidence for Evolution ....... 159
CHAPTER 24 - Evolution: Origin and History of Life ....... 165
CHAPTER 25 - The Microbial Kingdoms: Viruses, Monerans, Protists and Fungi ....... 171
CHAPTER 26 - The Plant Kingdom ....... 179
CHAPTER 27 - Animal Diversity ....... 185

## Part 6 - Ecology and Animal Behavior

CHAPTER 28 - Networks of Life: Ecology and the Biosphere ....... 193
CHAPTER 29 - Natural Order of Life: Ecosystems and Communities ....... 199
CHAPTER 30 - Life's Interactions: Communities ....... 207
CHAPTER 31 - Animal Behavior ....... 213
CHAPTER 32 - Population Dynamics ....... 219

## Appendix

Answers to Selected Study Guide Questions ....... 227

# Chapter 1

# Biology: The Study of Life

## Go Figure!

Figure 1-4: New discoveries in science often result from the introduction of new technologies. For example, in the 1880s molten materials, including wax, began to be used to infiltrate biological specimens so that thin slices could be made for examination under a microscope. This technology revolutionized the field. Which levels of biological organization do you think were unknown to science until new technological innovation revealed their presence?

Figure 1-5: This figure implies many historical relationships. What does it imply about the kingdom's historical sequence?

Which Kingdom is directly ancestral to members of Kingdom Fungi <u>and</u> the Animal Kingdom?

Which organisms are most closely related: slime molds and brown algae, brown algae and red algae, spiders and insects, ribbon worms and earthworms, or fish and sea stars (starfish)

Which type of organism has been on earth the longest, fern or conifer?

What does the caption imply about the relationship between number of species in a kingdom and the diversity of species within a kingdom?

Figure 1-10: The two jars in part "a" differ by more than one variable. Besides presence of adult flies and free exchange of air between the inside and outside of the jar, what other possible differences exist?

Would it be acceptable procedure to keep the wax sealed jar out of the sun so that the wax would not melt, while keeping the uncovered jar in the sun?

What would you conclude if after several days an open jar did <u>not</u> have maggots?

Figure 1-11: Why was the neck of the flask bent?

What was the purpose of heating the broth?

What would you conclude if no microbes grew after tipping the flask?

Figure 1-12: Assume you are applying this scheme to test the ability of some food additive to cause cancer. List all resources you can think of that could be used to complete the first step.

Using this scheme, what action would you take if the results of an experiment were different than your hypothesis predicted?

## Matching Questions

Write the letter of the phrase that best matches the numbered term on the left. Use each only once.

____ 1. biology
____ 2. cell
____ 3. unicellular
____ 4. metabolism
____ 5. energy
____ 6. gene
____ 7. development
____ 8. enzyme
____ 9. homeostasis
____ 10. stimuli
____ 11. spontaneous generation
____ 12. biosphere
____ 13. population
____ 14. adaptation
____ 15. hypothesis test
____ 16. experimental group
____ 17. variable
____ 18. theory
____ 19. hypothesis
____ 20. control group

a. all chemical reactions within an organism

b. a unit of hereditary instruction

c. the smallest organization level of life

d. a change in form during one individual's life time

e. the science of life

f. changes in the environment

g. the non-material property used by organisms to do the work of building and maintaining structure

h. in an experiment, those objects exposed to the variable being tested

i. used to describe an organism that consists of one cell

j. all members of a species living in the same habitat

k. a tentative, testable statement of cause and effect

l. includes every place on earth where living organisms are found

m. nonliving material producing living organisms

n. a collection of related hypotheses

o. a condition which can change

p. substance which increases the rate of a specific chemical reaction

q. maintenance of internal conditions within a limited range

r. increases the likelihood that an organism will reproduce

s. in an experiment, those objects not exposed to the variable being tested

t. an experiment

## Multiple Choice Questions

1. Which IS NOT used to differentiate living from nonliving things?
   a. high degree of complexity
   b. composed of one or more cells
   c. temperature fluctuates with environmental temperature
   d. produces offspring similar to parents
   e. responds to changes in the environment

2. Which IS NOT used to differentiate living from nonliving things?
   a. produces offspring similar to themselves
   b. inherits genes from parents
   c. grows in size and changes in appearance
   d. uses sunlight directly as a source of energy
   e. maintains homeostasis

3. The genetic instructions inherited from our parents consists of a large collection of
   a. preprogrammed behavioral responses.
   b. preformed body parts.
   c. miniature internal organs.
   d. genes.
   e. cell organelles.

4. The genetic instructions inherited from our parents are chemically composed of
   a. enzymes.
   b. energy.
   c. preformed body parts.
   d. DNA.
   e. cell organelles.

5. As a tadpole transforms into an adult frog, there is very little if any change in size, but there is a great deal in its external appearance. This modification is an example of
   a. reproduction.
   b. growth.
   c. development.
   d. metabolism.
   e. homeostasis.

6. In most species, as a tadpole transforms into an adult frog, its diet switches from plants to animals. This switch is accompanied by a change in the type of enzymes present in the digestive tract. This modification is an example of
   a. reproduction.
   b. growth.
   c. development.
   d. metabolism.
   e. homeostasis.

7. Organelles are composed of
   a. cells.
   b. other organelles and cytoplasm.
   c. molecules.
   d. populations.
   e. communities.

8. Cells are composed of
   a. cells.
   b. organelles and cytoplasm.
   c. molecules.
   d. populations.
   e. communities.

9. Which is the lowest level of organization capable of performing "higher" biological functions including sight?
   a. cell
   b. population
   c. multicellular organism
   d. atoms
   e. community

10. Which is the lowest level of organization capable of evolution?
    a. cell
    b. population
    c. multicellular organism
    d. atom
    e. community

11. A newly discovered species is given a "scientific name" composed of its
    a. genus and specific epithet.
    b. family and specific epithet.
    c. order and class names.
    d. family and genus names.
    e. family and order names.

12. The system of classification in use today assigns species into taxonomic categories based upon
    a. an alphabetical arrangement of scientific names.
    b. alphabetical arrangement of species names.
    c. evolutionary relationships between species.
    d. similarity in habitat.
    e. ability to maintain homeostasis.

13. Two organisms live in the same place, but are members of different populations; this implies that they are
    a. different species.
    b. members of different ecosystems.
    c. the same species.
    d. members of different communities.
    e. in the Kingdom Protista.

14. Two organisms live in different places, but have the same scientific name; this implies that they are
    a. different species.
    b. members of different families.
    c. members of different genera.
    d. members of different communities.
    e. in the Kingdom Protista.

15. Natural selection acts
    a. in a random manner upon all individuals in a population.
    b. upon variability with a population.
    c. equally upon all individuals in a population.
    d. by increasing the reproductive rate for an entire population.
    e. by decreasing the reproductive rate for an entire population.

16. Darwin proposed that, since all species produce more individuals than can survive,
    a. individuals in the population are variable.
    b. some individuals are better adapted than others.
    c. a population changes over time.
    d. some individuals in the population will not survive.
    e. offspring inherit the traits of their parents.

17. An well-designed experiment may attempt to
    a. establish blame.
    b. establish cause and effect.
    c. repeat all observations collected up to that point.
    d. establish cause and effect or repeat all observations collected up to that point.
    e. make predictions.

18. Which DOES NOT describe a well-designed experiment?
    a. It is designed to test a hypothesis.
    b. It can conclusively demonstrate cause and effect if there is a correlation between events.
    c. The hypothesis must be falsifiable--that is, have the potential to be shown incorrect if it is incorrect
    d. The sample is large enough to minimize the effect of chance on results.
    e. The control and experimental groups must differ in no more or less than one way.

*NOTE: Answers for questions 19 through 24 are provided in the Appendix. Questions 25 through 30 constitute a similar sequence, but the answers are not given.*

Use the following information for questions 19-24.

A pharmaceutical company has developed a new drug. In preliminary tests, the drug reduced high blood pressure in laboratory animals when injected under the skin. A second experiment was designed to more thoroughly test the drug's effectiveness. One thousand male rabbits were tested. Five hundred received the drug through injection (Group "A") and five hundred received nothing (Group "B"). All rabbits were the same breed, fed the same type and amount of food, and housed under identical conditions. After the drug was administered, Group "A" had elevated blood pressure for approximately one hour, then it returned to normal. The blood pressure of Group "B" remained unchanged and normal.

19. Which group of rabbits acted as a control?
    a Group "A"
    b. Group "B"
    c. neither group
    d. all 1000 rabbits
    e. half of group "A" and half of group "B"

20. Were the control and experimental groups adequately matched?
    a. Yes, because only male rabbits were used.
    b. Yes, because they were all the same breed, fed the same type and amount of food, and housed under identical conditions.
    c. No, because no female rabbits were used.
    d. No, because Group "B" did not receive an injection.
    e. No, because only Group "A" received the drug.

21. Assuming an adequate control was used, which hypothesis could be tested using this experimental design?
    a. The drug lowers blood pressure in some breeds of male rabbits but not in others.
    b. The drug does not lower blood pressure in male rabbits kept in the dark.
    c. The drug lowers blood pressure in male rabbits.
    d. The drug does not lower blood pressure when taken orally.
    e. The drug lowers blood pressure when taken orally.

22. Assuming an adequate control was used, which hypothesis could be tested using this experimental design?
    a. The drug does not lower blood pressure in overweight rabbits.
    b. The drug does not lower blood pressure in male rabbits.
    c. The drug lowers blood pressure in male rabbits housed in cages smaller than 0.5 X 0.5 X 0.5 meters.
    d. The drug lowers blood pressure in Group "B" rabbits
    e. The drug does not lower blood pressure in Group "B" rabbits.

23. Why were the rabbits "...of the same breed, were fed the same type and amount of food, and were housed under identical conditions."?
    a. to simplify their care and maintenance during the experiment
    b. to standardize the test
    c. to reduce possible psychological bias
    d. to reduce the number of variables
    e. to assure the rabbits have a normal heart rate

24. As described, how many variables are there between the control and experimental groups?
    a. 1
    b. 2
    c. 3
    d. 4
    e. 5

Use the following information for questions 25-30.

Melanoma is a type of cancer that produces a black-pigmented skin tumor. Its incidence is on the rise. For this and other reasons, a great deal of research is concentrating on this type of cancer. It has been hypothesized that injections of a vaccine containing chemical substances shed by melanoma cells grown in a research lab might stimulate the body's immune system (its natural defense system), thereby helping to reduce the tumor's growth. To test this hypothesis, one hundred melanoma patients were given the experimental vaccine by injection (Group "A"), and one hundred melanoma patients were given an injection of a substance known to have no physiological effect (Group "B"). All subjects were matched according to age, gender, and diet, and all reside in the United States. Group "A" subjects were from the midwest and Group "B" from the southeast. After the vaccine was administered, Group "A" showed a reduced rate of tumor growth. Tumor growth in Group "B" remained unchanged.

25. Which was the experimental group?
    a. Group "A"
    b. Group "B"
    c. neither group
    d. all 100 patients
    e. half of group "A" and half of group "B"

26. Were the control and experimental groups adequately matched?
    a. Yes, because the two groups were from different geographic regions.
    b. Yes, because they were matched according to age, gender, and diet.
    c. No, because both groups had melanoma.
    d. No, because Group "B" did not have melanoma.
    e. No, because they were not matched for general life style and geographical area of residence.

27. Assuming an adequate control was used, which hypothesis could be tested using this experimental design?
    a. The vaccine reduces tumor growth in males but not in females.
    b. The vaccine does not reduce tumor growth in subjects who smoke.
    c. The vaccine reduces tumor growth.
    d. The vaccine does not reduce tumor growth when applied to the skin.
    e. The vaccine reduces tumor growth when applied to the skin.

28. Assuming an adequate control was used, which hypothesis could be tested using this experimental design?
    a. The vaccine does not reduce tumor growth in people who smoke.
    b. The vaccine does not reduce tumor growth.
    c. The vaccine reduces tumor growth in people who live in rural areas.
    d. The vaccine reduces tumor growth in subjects in Group "B".
    e. The vaccine reduces tumor growth in Group "B".

29. Why were the subjects "...matched according to age, gender, and diet"?
    a. to simplify their care and maintenance during the experiment
    b. to reduce the number of variables
    c. to reduce possible psychological bias
    d. to standardize the test
    e. to assure the tumors have a normal growth rate

30. The control group received "...an injection of a substance known to have no physiological effect." What purpose did it serve?
    a. It simplified their care and maintenance during the experiment.
    b. It standardized the test.
    c. It reduced possible psychological bias.
    d. By general agreement, researchers always inject the control group.
    e. It reduced the number of variables.

## Concept Map Construction

Construct a concept map for each group of terms. Be sure to include appropriate connector phrases. You may add other terms as necessary and use terms in the plural or singular form.

1. biology, cell, unicellular, energy, reproduction
2. stimulus, homeostasis, atom, cell, population
3. experiment, variable, experimental group, hypothesis, observation

# Chapter 2

# The Chemical Basis of Life

## Go Figure!

Figure 2-4: The number in the center of each atom represents the number of positive charges in the atom. What is this number called?

Where are the positive charges located?

Which subatomic particle carries a positive charge?

Which subatomic particle is *not* represented in these atomic structure diagrams?

Figure 2-5: Movement of electrons from one orbital to another results in an increase of potential energy of the electron and it takes energy to do this. A similar situation occurs when a ball is rolled up a hill. An energy input is required to move the ball against the pull of gravity. Some of this energy is stored as potential energy. If the ball does not reach a position of stability (*i.e.,* resting on a flat spot or lodged behind a rock) when the energy input is removed, the ball will spontaneously begin rolling back down the hill all on its own. The energy for this movement comes from the stored potential energy and it will roll until it reaches a position of stability. The same applies to the electron. As long as the energy source is being used, the electron can remain in a less stable, higher energy orbital in the atom. But as soon as the energy is removed, the electron drops to a more stable, lower energy orbital. The potential energy is then converted to some other form, such as heat or light.

Figure 2-6: Notice the two different representations of each molecule. The upper drawing shows atomic nuclei and their electrons. The lower uses elemental symbols.

How many electrons must be shared to get a covalent bond?

What does each line represent between the letters?

Figure 2-7: What is the maximum number of hydrogen bonds a water molecule can form at any one instant?

Figure 2-10: Which is a more stable (*i.e.,* less likely to change) arrangement of molecules: those in the upper picture or those in the lower picture? Why?

Figure 2-11: Which is more acidic: coffee or acetic acid? By what factor?

Which is more acidic: coffee or baking soda? By what factor?

Which is more acidic: soap or baking soda? By what factor?

Figure 2-13: What happens to the "carrier" after it has delivered the monomer to the growing polymer?

Figure 2-14: What is the chemical formula for glucose?

Why do you suppose chemists use different styles to represent the same molecules?

Figure 2-18: What difference in the third fatty acid chain might be responsible for its bend?

Figure 2-19: Why is water represented only as a layer on each surface of the membrane?

Given the physical properties of the membrane and water molecules, suggest ways in which water might pass through the membrane.

Figure 2-22: What kind of bond is a peptide bond?

Which atoms are removed during peptide bond formation?

Figure 2-23: Notice that each level of organization in a protein retains lower ones. The primary, secondary, and tertiary structure of the polypeptide subunits is still visible in the drawings of secondary, tertiary, and quaternary structure, respectively. Find each of these.

## Matching Questions

Write the letter of the phrase that best matches the numbered term on the left. Use each only once.

____ 1. element
____ 2. electron
____ 3. isotope
____ 4. atomic number
____ 5. orbital
____ 6. nonpolar compound
____ 7. hydrogen bond
____ 8. hydrophilic molecule
____ 9. solvent
____10. acidic solution
____11. buffer
____12. organic molecule
____13. polymer
____14. polyunsaturated fat
____15. R group
____16. hydrolysis
____17. peptide bond
____18. adenosine triphosphate (ATP)
____19. monosaccharide
____20. amino acid
____21. denaturation

a. a simple sugar
b. a chemical reaction in which water splits a bond
c. loss of tertiary structure by a protein
d. negatively charged subatomic particle
e. the location of electrons in an atom
f. this part makes amino acids different
g. this has a pH less than 7.0
h. a substance that dissolves another substance
i. general term for molecules with an equal distribution of charges within them
j. it differs in its atomic mass from other atoms of the same element
k. this often contains a polar covalent bond
l. a molecule with many double bonds in it
m. a chemical that resists changes in pH by combining with $H^+$ or $OH^-$ ions
n. a nucleotide used as an energy source
o. a fundamental form of matter
p. the subunit of protein
q. a chemical bond between amino acids
r. a general term applied to a molecule that contains carbon
s. this equals the number of protons in the atom of an element
t. a large molecule made up of many subunits
u. a weak attraction between polar molecules

## Multiple Choice Questions

1. Which symbol is used to designate the element carbon?
   a. C
   b. Cl
   c. Ca
   d. Fe
   e. K

2. Which symbol is used to designate the element nitrogen?
   a. H
   b. Na
   c. O
   d. N
   e. Mg

3. Which describes an atom?
   a. It is a solid sphere.
   b. The nucleus contains protons and neutrons.
   c. The number of protons equals the number of neutrons.
   d. The nucleus contains electrons.
   e. Protons and neutrons have the same charge.

4. Which describes an atom?
   a. Most of the atom is empty space.
   b. Protons and electrons have the same charge.
   c. Electrons are positively charged.
   d. Neutrons and electrons have the same mass.
   e. Protons and electrons have the same mass.

5. Which is a false statement?
   a. Atomic mass is sometimes referred to as "atomic weight".
   b. Atomic mass may differ among atoms of the same element.
   c. Electrons, neutrons, and protons contribute significantly to atomic mass.
   d. Isotopes of an element have a different number of neutrons.
   e. Radioactive isotopes are unstable and break apart spontaneously.

6. Which is a false statement?
   a. All atoms of the same element have the same atomic number.
   b. All atoms of the same element have the same number of protons.
   c. Electrons occupying shells closer to the nucleus have more energy than those farther away from the nucleus.
   d. Electron orbitals may contain a maximum of two electrons.
   e. A neutrally charged atom will have an equal number of electrons and protons.

7. The atomic number of phosphorus is 15. How many electrons are in each shell of an atom of phosphorus? (Begin with the number in the innermost shell.)
   a. 15:0
   b. 5:8:2
   c. 2:6:7
   d. 2:8:5
   e. 2:2:2:2:2:2:2:1

8. The atomic number of magnesium is 12. How many electrons are in each shell of an atom of magnesium? (Begin with the number in the innermost shell.)
   a. 2:2:8
   b. 2:4:6
   c. 6:4:2
   d. 2:2:2:2:2:2
   e. 2:8:2

9. Which DOES NOT describe covalent bonds?
   a. Electrons are shared between atoms.
   b. The resulting structure is referred to as an ion.
   c. Electrons travel around the nuclei of both atoms.
   d. A double bond involves two pair of electrons.
   e. Some atoms can form more than one covalent bond.

10. Which DOES NOT describe ionic bonds?
   a. Ionic bonds form between charged atoms.
   b. Ionic bonds require complete transfer of an electron from one atom to another.
   c. Some atoms can form more than one ionic bond.
   d. Ionic bonds always involve pairs of electrons.
   e. Ionic bonds are one example of noncovalent bonds.

11. How many electrons must be gained by an atom of sulfur (atomic number = 16) in order to fill the outer electron shell?
   a. 1
   b. 2
   c. 3
   d. 4
   e. 5

12. How many electrons must an atom of fluorine (atomic number = 9) gain in its outer shell in order to maximize its stability?
   a. 1
   b. 2
   c. 3
   d. 4
   e. 5

13. What is the maximum number of covalent bonds an atom of phosphorus (atomic number = 15) can form?
   a. 1
   b. 2
   c. 3
   d. 4
   e. 5

14. What is the maximum number of covalent bonds an atom of silicon (atomic number = 14) can form?
   a. 1
   b. 2
   c. 3
   d. 4
   e. 5

15. Which DOES NOT describe polar molecules?
    a. They have regions of electrical charge within them.
    b. They result from unequal sharing of electrons in a covalent bond.
    c. They often participate in hydrogen bonds.
    d. The number of electrons exceeds the number of protons in the molecule.
    e. They are hydrophilic.

16. Which describes nonpolar molecules?
    a. They are ions.
    b. They are hydrophobic.
    c. They have regions of electrical charge within them.
    d. The number of electrons exceeds the number of protons in the molecule.
    e. They often participate in hydrogen bonds.

17. Which describes an acid?
    a. It is a substance that lowers the pH of a solution.
    b. If it is a strong acid, it will not give up its proton very easily.
    c. It is a substance that accepts protons from water.
    d. It will not react with a buffer, but may react with water.
    e. It will not react with $OH^-$ ions, but may react with a buffer.

18. Which describes a base?
    a. It is a substance that lowers the pH of a solution.
    b. It binds readily to $H^+$ ions.
    c. It binds readily to $OH^-$ ions.
    d. Water never acts as a base.
    e. Addition of a base would tend to make a solution less alkaline.

19. Which is an organic molecule?
    a. $C_3H_8O_3$
    b. $NO_2$
    c. $H_2O$
    d. $H_2SO_4$
    e. $O_2$

20. Which is an organic molecule?
    a. HCl
    b. $H_2S$
    c. $H_2$
    d. $CH_3COCOOH$
    e. NaCl

21. Which DOES NOT describe organic molecules?
    a. Organic molecules contain the element carbon.
    b. Organic molecules may be linear, branched, or cyclic.
    c. Organic molecules must have come from a living organism.
    d. Organic molecules may contain functional groups that give them special properties.
    e. Organic molecules may be polar or nonpolar.

22. Which DOES NOT describe biochemicals?
    a. The carbon chain may be linear, branched, or cyclic.
    b. They must have come from a living organism.
    c. They may contain functional groups that give them special properties.
    d. They may be polar or nonpolar.
    e. They are usually simple, small molecules.

23. Which DOES NOT describe biological macromolecules?
    a. They are often very small in size.
    b. They are often polymers.
    c. Their synthesis results from condensation reactions.
    d. Carrier molecules are removed from monomers prior to assembly of certain macromolecules.
    e. They are organic molecules.

24. Which DOES NOT describe biological macromolecules?
    a. Hydrolysis produces monomers from polymers.
    b. There are four basic families of of biological macromolecules.
    c. Biochemical families from different organisms are very different in basic structure.
    d. Diversity within a biochemical family occurs at the level of monomer sequences.
    e. Each biochemical family has basic functions.

25. Which is a true statement about carbohydrates?
    a. Glycogen is a structural polysaccharide in animals.
    b. Starch is a polysaccharide used for energy storage in plants.
    c. Chitin is a structural polysaccharide in most plants.
    d. Cellulose is easily digested by most animals.
    e. A peptide bond joins two monosaccharides into a disaccharide.

26. Which is a true statement about carbohydrates?
    a. Carbohydrates function as energy storage and structural molecules.
    b. Carbohydrates are typically hydrophobic.
    c. Carbohydrates contain more energy than an equivalent weight of fat.
    d. Monosaccharides function as hormones in many organisms.
    e. Nucleotides are the subunits of polysaccharides.

27. Which molecule is used for short term energy storage?
    a. fat
    b. chitin
    c. glycerol
    d. glycogen
    e. cellulose

28. Which molecule is used for long term energy storage?
    a. fat
    b. chitin
    c. glycerol
    d. glycogen
    e. cellulose

29. Which describes lipids?
    a. Lipids are soluble in water.
    b. Lipids are hydrophilic.
    c. Lipids are nonpolar.
    d. Lipids are inorganic molecules.
    e. Lipids include molecules like polypeptides.

30. Which is a true statement?
    a. Fats contain a phosphate group.
    b. Steroids are made of glycerol and three fatty acids.
    c. Phospholipids resemble steroids in molecular structure.
    d. Unsaturated fats tend to be liquid at room temperature.
    e. Fats have a hydrophilic and a hydrophobic portion.

31. Which class of molecule is the major component of cell membranes?
    a. phospholipid
    b. cellulose
    c. wax
    d. glycogen
    e. monosaccharide

32. Which molecule is used to make plant cell walls?
    a. disaccharide
    b. polypeptide
    c. cellulose
    d. phospholipid
    e. chitin

33. Which is a true statement about proteins?
    a. Each organism has relatively few proteins.
    b. Proteins are the primary form of energy storage in most organisms.
    c. Proteins are fatty acid polymers.
    d. Proteins from all organisms contain the same 20 amino acids.
    e. All proteins have the same three dimensional structure.

34. Which DOES NOT describe proteins?
    a. The sequence of amino acids determines the primary structure of a protein.
    b. The secondary structure of a protein may be in the form of a helix, pleated sheet, or random coil.
    c. Physical properties of the R groups in the protein help determine tertiary structure.
    d. Denaturation of a protein involves loss of primary structure.
    e. The effects of denaturation may not be permanent.

35. Which molecule is responsible for storing hereditary information?
    a. deoxyribonucleic acid (DNA)
    b. steroids
    c. ribonucleic acid (RNA)
    d. monosaccharides
    e. polypeptides

36. Enzymes are
    a. lipids.
    b. proteins.
    c. carbohydrates.
    d. nucleic acids.
    e. inorganic molecules.

## Concept Map Construction

Construct a concept map for each group of terms. Be sure to include appropriate connector phrases. You may add other terms as necessary and use terms in the singular or plural form.

1. scientific method, element, sulfur, subatomic particle, isotope
2. functional group, amino acid, polymer, lipid, hydrogen bond
3. R group, biochemical, atomic nucleus, ATP, nucleotide

# Chapter 3

# Cell Structure and Function

## Go Figure!

Figure 3-2: Which differences between prokaryotic and eukaryotic cells are visible in this figure?

Figure 3-3: What do the reddish globular objects in this model represent?

Why do some go completely through the membrane while others are only on the surface?

Figure 3-4: Which organelles depicted in this figure are present only in plant cells?

Which are present only in animal cells?

Which are common to both?

Figure 3-5: Why are no chromosomes visible in the nucleus?

Figure 3-6: What is the purpose of the genetic material becoming tightly coiled prior to cell division?

Notice the chromosome has two parts. What process has occurred to produce those two parts?

Figure 3-7: Are the ribosomes on the cytoplasmic side or the internal side of the endoplasmic reticulum membranes?

Figure 3-8: Why are Golgi complexes "clustered near the area where new cell wall is being manufactured"? (Check Figure 3-9 for a clue.)

Figure 3-11: How are chloroplasts and mitochondria similar?

How are they different?

Figure 3-15: Why are membrane channels probably unnecessary for nonionic compounds?

Figure 3-16: Does any water enter or leave the cell when in an isotonic solution?

Does any water enter the cell when in a hypertonic solution?

Does any water leave the cell when in a hypotonic solution?

Figure 3-18: Why must a particle in a vacuole be digested before it can enter a cell's cytoplasm?

## Matching Questions

Write the letter of the phrase that best matches the numbered term on the left. Use each only once.

____ 1. smooth endoplasmic reticulum

____ 2. prokaryotic cell

____ 3. Golgi complex

____ 4. cytoplasm

____ 5. lysosome

____ 6. cilium

____ 7. phospholipid bilayer

____ 8. plant cell wall

____ 9. vacuole

____10. ribosome

____11. nuclear envelope

____12. nucleolus

____13. selective permeability

____14. carrier protein

____15. osmosis

____16. simple diffusion

____17. facilitated diffusion

____18. active transport

____19. hypertonic

____20. hypotonic

____21. exocytosis

a. solute concentration outside cell is more than the cellular solute concentration

b. produces secretory vesicles

c. a primitive type of cell

d. especially prominent in mature plant cells

e. the basic structure of membrane

f. the final step in the secretory pathway

g. diffusion of water across a selectively permeable membrane

h. consists of a double membrane

i. carrier-assisted movement against a concentration gradient

j. site of ribosome formation

k. the cellular region outside the nucleus

l. organelle with hydrolytic enzymes

m. involved in the synthesis of steroid hormones

n. movement down a concentration gradient

o. carrier-assisted movement along a concentration gradient

p. solute concentration outside cell is less than cellular solute concentration

q. composed of cellulose

r. a membrane component necessary for active transport

s. site of protein synthesis

t. a short projection used for movement

u. a property of plasma membranes

## Multiple Choice Questions

1. Which is a true statement?
   a. Cells were discovered by Anton van Leeuwenhoek in the 1600s.
   b. Cells were first seen in pond water.
   c. The discovery of cells required the invention of the light microscope.
   d. The name "cell" was first used by Theodor Schwann.
   e. Cells are found only in plants.

2. Which DOES NOT describe the cell theory?
   a. All living things are made of cells.
   b. Cells may develop from nonliving matter.
   c. Cells are the smallest organizational unit that have all the properties of life.
   d. Theodor Schwann and Matthias Schleiden contributed to development of the cell theory.
   e. Rudolf Virchow contributed to the development of the cell theory.

3. Which is a true statement concerning prokaryotic cells?
   a. Prokaryotic cells are generally larger than eukaryotic cells.
   b. Prokaryotic cells have a well-defined nucleus.
   c. Prokaryotic cells are found in bacteria.
   d. Prokaryotic cells evolved from eukaryotic cells.
   e. Prokaryotic cells lack ribosomes.

4. Which is a true statement concerning eukaryotic cells?
   a. Eukaryotic cells are capable of more complex activities than prokaryotic cells.
   b. Eukaryotic cells never have a cell wall.
   c. Eukaryotic cells lack membranous organelles.
   d. Eukaryotic cells lack ribosomes.
   e. Eukaryotic cells are simpler in structure than prokaryotic cells.

5. Which DOES NOT describe the nucleus of eukaryotic cells?
   a. It contains most of the DNA of the cell.
   b. The nuclear envelope consists of a single membrane.
   c. Chromosomes are visible only when a cell is dividing.
   d. It contains one or more nucleoli.
   e. Nuclear pores in the envelope allow passage of materials between the nucleus and the cytoplasm.

6. Which DOES NOT describe endoplasmic reticulum?
   a. Rough endoplasmic reticulum is associated with ribosomes.
   b. Rough endoplasmic reticulum is most abundant in secretory cells.
   c. Rough endoplasmic reticulum is part of the secretory pathway.
   d. Smooth endoplasmic reticulum may destroy toxic substances.
   e. Smooth endoplasmic reticulum is involved in protein synthesis.

7. Which organelle IS NOT made of membrane?
   a. endoplasmic reticulum
   b. chloroplast
   c. Golgi complex
   d. lysosome
   e. ribosome

8. Which IS NOT made of protein?
   a. microtubule
   b. microfilament
   c. intermediate filament
   d. vesicle
   e. cytoskeleton

9. Which is the proper sequence of the secretory pathway?
   a. endoplasmic reticulum, Golgi complex, ribosome
   b. Golgi complex, endoplasmic reticulum, chloroplast
   c. mitochondrion, vesicle, cilia
   d. endoplasmic reticulum, Golgi complex, vesicle
   e. Golgi complex, endoplasmic reticulum, mitochondrion

10. Which process is involved in secretion?
   a. exocytosis
   b. hydrolysis
   c. endocytosis
   d. ciliary action
   e. phagocytosis

11. Which DOES NOT describe lysosomes?
   a. They are products of the Golgi complex.
   b. They contain hydrolytic enzymes.
   c. They are active during destruction of a cell's own worn-out organelles.
   d. They are responsible for certain diseases.
   e. They are the site of aerobic respiration.

12. Which DOES NOT describe vacuoles?
   a. They are prominent in many mature plant cells.
   b. They are the site of photosynthesis.
   c. They are made of a single membrane.
   d. They help maintain internal water pressure.
   e. They accumulate wastes from the cell.

13. Which DOES NOT describe chloroplasts?
   a. They perform photosynthesis.
   b. They contain DNA and ribosomes.
   c. They are made of a double membrane.
   d. They are found in plants and fungi.
   e. Thylakoid membranes contain chlorophyll.

14. Which DOES NOT describe mitochondria?
   a. They are made of a double membrane.
   b. They contain DNA and ribosomes.
   c. They are most abundant in inactive cells.
   d. They are found in plants and fungi.
   e. They are involved in ATP production.

15. Which describes the endosymbiosis hypothesis?
   a. It is used to explain the origin of all membranous organelles.
   b. It is used to explain the origin of ribosomes.
   c. It is used to explain the origin of prokaryotic cells.
   d. It is used to explain the origin of chloroplasts.
   e. It is used to explain the origin of Golgi complexes.

16. Which IS NOT evidence supporting the endosymbiosis hypothesis?
   a. Mitochondria have their own DNA similar to prokaryotic DNA.
   b. Chloroplasts have their own ribosomes similar to prokaryotic ribosomes.
   c. Endosymbiotic relationships have been observed in modern prokaryotes.
   d. The inner membrane of mitochondria resembles the plasma membrane of respiratory bacteria.
   e. The outer membrane of chloroplasts resembles the plasma membrane of photosynthetic bacteria.

17. Which part of the plasma membrane makes it permeable to lipid soluble substances?
    a. phospholipids
    b. carrier proteins
    c. carbohydrates
    d. nucleic acids
    e. water

18. Which makes the plasma membrane selectively permeable to lipid insoluble substances?
    a. steroids
    b. carbohydrates
    c. nucleic acids
    d. carrier proteins
    e. water

19. Which DOES NOT describe simple diffusion across a plasma membrane?
    a. The membrane must be impermeable to the diffusing substance.
    b. Diffusion occurs down a concentration gradient.
    c. Diffusion of hydrophobic molecules may occur through the phospholipid bilayer.
    d. Diffusion may occur through protein channels.
    e. The diffusing particle may be lipid soluble or lipid insoluble.

20. Which DOES NOT describe simple diffusion across a plasma membrane?
    a. Net movement occurs from a high concentration to a low concentration.
    b. Diffusion may be regulated by gated channels.
    c. Particles only move in one direction across the membrane.
    d. Diffusion of ions may occur through protein channels.
    e. Particle movement continues after the concentration is equal on both sides of the membrane.

21. How are facilitated diffusion and active transport similar?
    a. Both require ATP hydrolysis to supply the energy for transport.
    b. Both are responsible for the transport of water across the membrane.
    c. Both use membrane bound carrier proteins.
    d. Both work along a concentration gradient.
    e. Both work against a concentration gradient.

22. How are facilitated diffusion and active transport different?
    a. Facilitated diffusion requires ATP to supply the energy for transport; active transport does not.
    b. Facilitated diffusion transports water; active transport does not.
    c. Facilitated diffusion works against a concentration gradient; active transport works down a concentration gradient.
    d. Facilitated diffusion works down a concentration gradient; active transport works against a concentration gradient.
    e. Facilitated diffusion uses a membrane bound carrier protein; active transport does not.

23. A red blood cell has a solute concentration of 0.9%. What will happen if it is placed in a 0.8% salt solution?
    a. The red blood cell will shrink if its membrane is permeable to salt and water.
    b. The red blood cell will shrink if its membrane is impermeable to salt and permeable to water.
    c. The red blood cell will swell and probably burst if its membrane is permeable to salt and water.
    d. The red blood cell will swell and probably burst if its membrane is impermeable to salt and permeable to water.
    e. The red blood cell will shrink regardless of its permeability to salt and water.

24. A red blood cell has a solute concentration of 0.9%. What will happen if it is placed in an 8.0% salt solution?
    a. The red blood cell will shrink if its membrane is permeable to salt and water.
    b. The red blood cell will shrink if its membrane is impermeable to salt and permeable to water.
    c. The red blood cell will swell and probably burst if its membrane is permeable to salt and water.
    d. The red blood cell will swell and probably burst if its membrane is impermeable to salt and permeable to water.
    e. The red blood cell will shrink regardless of its permeability to salt and water.

25. What will happen to a plant cell placed in a hypotonic sugar solution?
    a. The plant cell will swell and probably burst if its membrane is permeable to sugar and water.
    b. The plant cell will swell and probably burst if its membrane is impermeable to sugar and permeable to water.
    c. The plant cell will shrink if its membrane is permeable to sugar and water.
    d. The plant cell will shrink if its membrane is impermeable to sugar and permeable to water.
    e. The pressure in the cell will increase.

26. What will happen to a plant cell placed in a hypertonic sugar solution?
    a. The plant cell will shrink if its membrane is permeable to sugar and water.
    b. The plant cell will shrink if its membrane is impermeable to sugar and permeable to water.
    d. The plant cell will swell and probably burst if its membrane is permeable to sugar and water.
    c. The plant cell will swell and probably burst if its membrane is impermeable to sugar and permeable to water.
    e. The pressure in the cell will increase.

27. Some white blood cells protect us from infectious bacteria by transporting them into the white blood cell's cytoplasm. By what process does this occur?
    a. simple diffusion
    b. active transport
    c. phagocytosis
    d. facilitated diffusion
    e. exocytosis

28. Once the bacterium is taken in by the white blood cell, it is digested by hydrolytic enzymes. What organelle carries these enzymes?
    a. mitochondrion
    b. ribosome
    c. lysosome
    d. rough endoplasmic reticulum
    e. smooth endoplasmic reticulum

## Concept Map Construction

Construct a concept map for each group of terms. Be sure to include appropriate connector phrases. You may add other terms as necessary and use terms in the singular or plural form.

1. rough endoplasmic reticulum, protein, phospholipid, microfilament, cell wall
2. Golgi complex, carbon, hydrolysis, mitochondrion, endosymbiotic hypothesis
3. endocytosis, membrane invagination hypothesis, active transport, ATP, concentration gradient

# Chapter 4

# The Metabolism of Life

## Go Figure!

Figure 4-1: Which laws of thermodynamics are illustrated by the energy transfers shown in this figure?

Figure 4-2: How could this house be restored to "prime real estate?"

Would restoration of the house violate the second law of thermodynamics since it represents an increase in order? Justify your answer.

Figure 4-4: Which reaction (exergonic or endergonic) could represent your lunch being digested?

Which reaction could represent new muscle tissue being made?

Figure 4-5: Do the energy levels of the reactants and products differ with or without enzyme?

Is the reaction illustrated exergonic or endergonic?

Figure 4-6: Is the conversion of Substrates A and B to Product C in Figure 4-6b catabolic or anabolic?

Is the conversion of Substrate C to Products A and B catabolic or anabolic?

Figure 4-7: Methanol ($CH_3OH$) and methane ($CH_4$) molecules each contain four hydrogen atoms and one carbon atom. Why is the carbon of methanol considered to be in a more oxidized state?

Formic acid (HCOOH) and carbon dioxide ($CO_2$) molecules each contain two oxygen atoms and one carbon atom. Why is the carbon of formic acid considered to be in a more reduced state?

What general principle about oxidation state can be derived from these examples?

Figure 4-8: What will activate Enzyme BC if it has been inhibited by Product E?

## Matching Questions

Write the letter of the phrase that best matches the numbered term on the left. Use each only once.

____ 1. cofactor

____ 2. endergonic reaction

____ 3. exergonic reaction

____ 4. activation energy

____ 5. substrate

____ 6. first law of thermodynamics

____ 7. catabolic pathway

____ 8. anabolic pathway

____ 9. oxidation reaction

____10. reduction reaction

____11. ATP

____12. active site

____13. NADPH

____14. entropy

____15. feedback inhibition

____16. enzyme

____17. coupled reactions

a. an inorganic chemical that assists an enzyme

b. a reactant that binds specifically to an enzyme's active site

c. an increase in product decreases its own production

d. a reaction in which energy is released

e. this reaction involves gaining an electron

f. a reaction in which the products have more energy than the reactants

g. ATP hydrolysis and an endergonic reaction, for example

h. the location of substrate binding

i. a measure of disorder

j. an organic catalyst

k. energy required to start a chemical reaction

l. "energy can neither be created nor destroyed"

m. series of reactions in which simple molecules are made into a complex one

n. reserves of this constitute "reducing power"

o. a nucleotide which supplies energy to chemical reactions

p. this reaction involves the loss of an electron

q. series of reactions in which a complex molecule is made into simpler ones

## Multiple Choice Questions

1. "The amount of energy in the universe is finite" is another way of stating the
   a. cell theory.
   b. endosymbiotic hypothesis.
   c. cell invagination hypothesis.
   d. first law of thermodynamics.
   e. second law of thermodynamics.

2. Which IS NOT consistent with the second law of thermodynamics?
   a. Entropy increases.
   b. The amount of usable energy decreases.
   c. Organisms use energy to maintain their organization.
   d. The chemical energy of food is used over and over as an organism lives it life.
   e. Events tend to occur from a higher to a lower energy state.

3. Which describes ATP?
   a. Its hydrolysis is an endergonic reaction.
   b. Its sugar is deoxyribose.
   c. Each molecule contains only two phosphates.
   d. Its hydrolysis is energetically favored in the cell.
   e. Its hydrolysis decreases entropy.

4. Which describes ATP?
   a. Its hydrolysis can be coupled to endergonic reactions of metabolism.
   b. It is a chemical used for long term energy storage.
   c. The amount of energy released during its hydrolysis is equal to or less than the amount of energy stored in a coupled endergonic reaction.
   d. One product of its hydrolysis is water.
   e. Cells maintain large concentrations of it.

5. Which describes enzymes?
   a. They are carbohydrates.
   b. They catalyze chemical reactions.
   c. Each type reacts with a variety of substrates.
   d. Each type is found in large concentrations in the cell.
   e. They increase the activation energy of a reaction.

6. Which describes enzymes?
   a. They bind specific substrates at their active site.
   b. Each may be used only once since they are chemically changed at the end of a reaction.
   c. Enzymes are unaffected by changes in pH.
   d. Since enzymes are chemicals, an increase in temperature always increases the rate of reaction.
   e. Coenzymes are inorganic molecules that assist enzymes in their function.

7. Which DOES NOT describe a catabolic pathway?
   a. Energy is released.
   b. The pathway is exergonic.
   c. It is a biosynthetic pathway.
   d. The reactants are more complex than the products.
   e. All the steps are catalyzed by different enzymes.

8. Which DOES NOT describe an anabolic pathway?
   a. The products are more complex than the reactants.
   b. Although energy is stored, entropy increases.
   c. Some intermediate reactions involve ATP hydrolysis.
   d. It never involves electron transfer.
   e. Each metabolic intermediate is the substrate for a different enzyme.

9. Which describes the situation when an atom becomes oxidized?
   a. The atom has gained an electron.
   b. The atom has gained energy.
   c. The atom may have combined with a hydrogen atom.
   d. The atom may have combined with an oxygen atom.
   e. As the atom becomes oxidized, it is not reacting with any other atom.

10. Which describes the situation when an atom becomes reduced?
    a. The atom becomes smaller.
    b. The atom may have combined with a hydrogen atom.
    c. The atom may have combined with an oxygen atom.
    d. As the atom becomes reduced, it is not reacting with any other atom.
    e. The oxidation state of the atom is unchanged.

11. How does covalent modification regulate enzyme activity?
    a. Often a phosphate group is added to a coenzyme to activate it.
    b. Often a phosphate group is removed from a cofactor to activate it.
    c. Often a phosphate group is added to the substrate so it doesn't fit the active site of the enzyme.
    d. Often a phosphate group is added to the enzyme so the active site won't bind the substrate.
    e. Often a phosphate group is removed from a coenzyme to inactivate it.

12. How does feedback inhibition regulate enzyme activity?
    a. The final product of a metabolic pathway activates the first enzyme in the pathway by binding to its active site.
    b. The final product of a metabolic pathway inactivates the first enzyme in the pathway by binding to its active site.
    c. The final product of a metabolic pathway activates the first enzyme in the pathway by binding to its feedback site.
    d. The final product of a metabolic pathway inactivates the first enzyme in the pathway by binding to its feedback site.
    e. The final product of a metabolic pathway inactivates the last enzyme in the pathway by binding to its feedback site.

13. Examine the reaction below.

$$\text{Pyruvate} + CO_2 + \text{ATP} + H_2O \longrightarrow \text{Oxaloacetate} + \text{ADP} + P_i$$

    Which is a true statement?
    a. The reaction is anabolic.
    b. The reaction is catabolic.

14. Examine the reaction below.

$$\text{Lactose} + H_2O \longrightarrow \text{Galactose} + \text{Glucose}$$

    Which is a true statement?
    a. The reaction is anabolic.
    b. The reaction is catabolic.

15. Examine the reaction below.

$$\text{Pyruvate} + CO_2 + \text{ATP} + H_2O \longrightarrow \text{Oxaloacetate} + \text{ADP} + P_i$$

    Which is a true statement?
    a. The conversion of pyruvate and $CO_2$ to oxaloacetate is endergonic.
    b. The conversion of pyruvate and $CO_2$ to oxaloacetate is exergonic.

16. Examine the reaction below.

1,3-biphosphoglycerate + ADP $\longrightarrow$ 3-phosphoglycerate + ATP

Which is a true statement?
a. The conversion of 1,3-biphosphoglycerate to 3-phosphoglycerate is endergonic.
b. The conversion of 1,3-biphosphoglycerate to 3-phosphoglycerate is exergonic.

17. The coenzyme NAD (nicotinamide adenine dinucleotide) is an electron carrier similar to NADP. With that in mind, examine the reaction below.

Glyceraldehyde-3-phosphate + $P_i$ + $NAD^+$ $\longrightarrow$ 1,3-bisphosphoglycerate + NADH + $H^+$

Which is a true statement?
a. Glyceraldehyde-3-phosphate is oxidized to become 1,3-bisphosphoglycerate.
b. Glyceraldehyde-3-phosphate is reduced to become 1,3-bisphosphoglycerate.

18. The coenzyme NAD (nicotinamide adenine dinucleotide) is an electron carrier similar to NADP. With that in mind, examine the reaction below.

Pyruvate + NADH $\longrightarrow$ Lactate + $NAD^+$

Which is a true statement?
a. Pyruvate becomes oxidized to form lactate.
b. Pyruvate becomes reduced to form lactate.

19. Examine the reaction below.

Pyruvate + $NH_4^+$ + NADH + $H^+$ $\longrightarrow$ L-alanine + $NAD^+$ + $H_2O$

Which is a true statement?
a. Pyruvate becomes oxidized to form L-alanine and the reaction is anabolic.
b. Pyruvate becomes reduced to form L-alanine and the reaction is anabolic.
c. Pyruvate becomes oxidized to form L-alanine and the reaction is catabolic.
d. Pyruvate becomes reduced to form L-alanine and the reaction is catabolic.

20. Examine the reaction below.

Glutamate + $NADP^+$ + $H_2O$ $\longrightarrow$ $\alpha$-ketoglutarate + $NH_4^+$ + NADPH + $H^+$

Which is a true statement?
a. Glutamate becomes oxidized to form $\alpha$-ketoglutarate and the reaction is anabolic.
b. Glutamate becomes reduced to form $\alpha$-ketoglutarate and the reaction is anabolic.
c. Glutamate becomes oxidized to form $\alpha$-ketoglutarate and the reaction is catabolic.
d. Glutamate becomes reduced to form $\alpha$-ketoglutarate and the reaction is catabolic.

## Concept Map Construction

Construct a concept map for each group of terms. Be sure to include appropriate connector phrases. You may add other terms as necessary and use terms in the singular or plural form.

1. energy, coenzyme, oxidation reaction, atomic nucleus, carbon
2. NAD, first law of thermodynamics, ribosome, mitochondrion, enzyme
3. feedback inhibition, exergonic reaction, coupled reactions, active site, carbohydrate

# Chapter 5

# Energy and Life: Photosynthesis

## Go Figure!

Figure 5-2: ATP and NADPH are hydrolyzed and oxidized respectively in the light-independent reactions. What happens to the resulting ADP and $NADP^+$?

Figure 5-3: Examine the absorption spectra of plant pigments. Why are chlorophylls green and carotenoids yellow, orange or red?

Figure 5-5: Of what adaptive value is having the inner membrane of the chloroplast so highly folded?

Figure 5-7: Experimental evidence indicates that four ATPs are produced per $O_2$ released during noncyclic photosynthesis. How many ATPs are produced per water molecule that undergoes photolysis?

Which chemical activities contribute to development of the proton gradient in noncyclic photophosphorylation?

Figure 5-8: Several activities generate the proton gradient during noncyclic photophosphorylation. Which of those activities are *not* involved in generating the proton gradient during cyclic photophosphorylation?

Figure 5-9: How many carbons (total) are in the two PGALs "drained" away to produce glucose?

How does this number compare to the original input of carbons in $CO_2$?

How many carbons are in the original six molecules of RuBP?

How does this compare to the number of carbons in the ten PGALs remaining after two have been removed to make glucose?

## Matching Questions

Write the letter of the phrase that best matches the numbered term on the left. Use each only once.

____ 1. chemosynthesis

____ 2. light-independent reactions

____ 3. $CO_2$ fixation

____ 4. RuBP carboxylase

____ 5. autotroph

____ 6. heterotroph

____ 7. ATP synthase

____ 8. chemiosmosis

____ 9. chlorophyll

____10. carotenoid

____11. P700

____12. P680

____13. cyclic photophosphorylation

____14. noncyclic photophosphorylation

____15. photolysis

____16. photorespiration

____17. stomate

____18. CAM plants

a. a primary function of the light-independent reactions

b. the enzyme that catalyzes ATP synthesis

c. an organism capable of photosynthesis or chemosynthesis

d. ATP and NADPH are produced

e. interference of photosynthesis by high $O_2$ concentration

f. autotrophic metabolism in which chemical (not light) energy is used

g. the enzyme that catalyzes carbon dioxide fixation in $C_3$ plants

h. yellow or orange plant pigment

i. an organism incapable of photosynthesis or chemosynthesis

j. carbon fixation and reduction is accomplished by these reactions

k. the process of splitting water

l. the reaction center of photosystem II

m. they fix $CO_2$ at night

n. a pore that allows $CO_2$ to enter the leaf

o. the only product is ATP

p. a green plant pigment

q. the reaction center of photosystem I

r. a model to describe the process of ATP formation

## Multiple Choice Questions

1. Which DOES NOT describe autotrophs?
   a. Most get energy from the sun.
   b. Some get energy from the oxidation of chemicals.
   c. They get their organic molecules from the soil.
   d. Most release $O_2$ as a waste product of their activity.
   e. They may be plants, bacteria, or protists.

2. Which DOES NOT describe autotrophs?
   a. They comprise the majority of species.
   b. Some perform chemosynthesis.
   c. Some perform photosynthesis.
   d. They convert $CO_2$ into organic molecules.
   e. They are the immediate energy source for many heterotrophs.

3. Which is a reactant in photosynthesis?
   a. carbohydrate ($CH_2O$)
   b. water ($H_2O$)
   c. oxygen ($O_2$)
   d. methane ($CH_4$)
   e. ammonia ($NH_3$)

4. Which is a product of photosynthesis?
   a. carbon dioxide ($CO_2$)
   b. light
   c. water ($H_2O$)
   d. oxygen ($O_2$)
   e. methane ($CH_4$)

5. Which compound supplies the carbon used during photosynthesis?
   a. carbon dioxide ($CO_2$)
   b. light
   c. water ($H_2O$)
   d. oxygen ($O_2$)
   e. methane ($CH_4$)

6. Oxygen produced during photosynthesis comes from which compound?
   a. carbon dioxide ($CO_2$)
   b. light
   c. water ($H_2O$)
   d. oxygen ($O_2$)
   e. methane ($CH_4$)

7. What is accomplished in the light-dependent reactions of photosynthesis?
   a. ATP is produced.
   b. NADPH is produced.
   c. Carbohydrate is produced.
   d. ATP and NADPH are produced.
   e. ATP, NADPH and carbohydrate are produced.

8. What is accomplished in the light-independent reactions of photosynthesis?
   a. ATP is produced.
   b. NADPH is produced.
   c. Carbohydrate is produced.
   d. ATP and NADPH are produced.
   e. ATP, NADPH and carbohydrate are produced.

9. Which DOES NOT describe antenna pigments?
   a. They may be carotenoids.
   b. They may be chlorophylls.
   c. They greatly outnumber the reaction center pigments.
   d. They convert light energy directly into chemical bond energy.
   e. They transfer light energy to a reaction center.

10. Which DOES NOT describe plant pigments?
    a. Chlorophylls contain magnesium.
    b. Chlorophylls absorb red and blue light.
    c. Chlorophylls are the primary plant pigment.
    d. Carotenoids absorb wavelengths not efficiently absorbed by chlorophylls.
    e. Carotenoids reflect green light.

11. Which DOES NOT describe Photosystem I?
    a. P700 is the reaction center.
    b. The antenna pigments include chlorophylls.
    c. The antenna pigments are embedded in the thylakoid membranes.
    d. There is a single Photosystem I in each chloroplast.
    e. Antenna pigments absorb light with wavelengths between 400 and 700 nm.

12. Which DOES NOT describe Photosystem II?
    a. P700 is the reaction center.
    b. The antenna pigments include carotenoids.
    c. The antenna pigments are embedded in the thylakoid membranes.
    d. There are many Photosystems II in each chloroplast.
    e. Antenna pigments absorb light with wavelengths between 400 and 700 nm.

13. Which DOES NOT describe Photosystem II?
    a. It is closely associated with oxygen production.
    b. It is responsible for increasing hydrogen ion concentration in the thylakoid lumen.
    c. It participates in photolysis.
    d. It receives electrons from water molecules.
    e. It transfers electrons directly to $NADP^+$.

14. Which DOES NOT describe Photosystem I?
    a. It receives electrons from Photosystem II via an electron transport system.
    b. It donates the electrons used in $NADP^+$ reduction.
    c. It is responsible for decreasing hydrogen ion concentration in the stroma of the chloroplast.
    d. It contains the ATP synthase involved in ATP production.
    e. P700 receives electrons from P680.

15. What is produced during noncyclic photophosphorylation?
    a. ATP
    b. NADPH
    c. $H_2O$
    d. ATP and NADPH
    e. ATP, NADPH and $H_2O$

16. What is produced during cyclic photophosphorylation?
    a. ATP
    b. NADPH
    c. $H_2O$
    d. ATP and NADPH
    e. ATP, NADPH and $H_2O$

17. The term "chemiosmosis" refers to the production of
    a. ATP.
    b. NADPH.
    c. carbohydrate.
    d. $O_2$.
    e. water.

18. The term "photophosphorylation" refers to the production of
    a. ATP.
    b. NADPH.
    c. carbohydrate.
    d. $O_2$.
    e. water.

19. Which DOES NOT contribute to the proton gradient created during the light dependent reactions of photosynthesis?
    a. impermeability to protons of the thylakoid membrane
    b. splitting of water
    c. reduction of $NADP^+$
    d. movement of protons associated with electron transport
    e. oxidation of NADPH

20. Which DOES NOT describe the ATP synthase in chloroplasts?
    a. It is not attached to a membrane.
    b. The stalked portion forms a proton channel.
    c. The active site has ADP and $P_i$ as substrates.
    d. It requires diffusing protons to drive synthesis.
    e. It is a structurally complex enzyme.

21. Which DOES NOT describe $C_3$ synthesis?
    a. The reactions are light dependent.
    b. $CO_2$ is attached to a five-carbon molecule.
    c. Ribulose biphosphate is involved in carbon dioxide fixation.
    d. NADPH from the light dependent reactions is a reactant.
    e. The pathway is cyclic.

22. Which DOES NOT describe $C_3$ synthesis?
    a. The first identifiable product of carbon dioxide fixation is a three-carbon compound.
    b. RuBP carboxylase is the enzyme of carbon dioxide fixation.
    c. ATP from the light dependent reactions is a reactant.
    d. The pathway is used by CAM and $C_4$ plants.
    e. Glucose is an intermediate of the Calvin-Benson Cycle.

23. Which describes $C_4$ plants?
    a. They are usually out-competed by $C_3$ plants in warm temperatures.
    b. They are adapted to a hot, dry environment.
    c. Carbon dioxide fixation occurs at night.
    d. Their carbon-fixing enzyme is less efficient at low $CO_2$ concentrations than the one used by $C_3$ plants.
    e. Carbon reduction occurs in mitochondria.

24. Which describes CAM plants?
    a. Carbon dioxide fixation occurs at night.
    b. The first identifiable product of carbon dioxide fixation is ribulose biphosphate.
    c. They comprise a majority of plants.
    d. They are found in cold, moist environments.
    e. Their carbon-fixing enzyme is less efficient at low $CO_2$ concentrations than the one used by $C_3$ plants.

## Concept Map Construction

Construct a concept map for each group of terms. Be sure to include appropriate connector phrases. You may add other terms as necessary and use terms in the singular or plural form.

1. chemosynthesis, eukaryote, NADPH, RuBP carboxylase, mitochondrion
2. photosynthesis, ATP, P700, water, active site
3. ATP synthase, light dependent reactions, CAM plant, chlorophyll, diffusion

# Chapter 6

# Energy and Life: Respiration and Fermentation

## Go Figure!

Figure 6-2: Where do the two 1,3 Diphosphoglycerates (DPG) come from?

Where do the phosphates on 1,3-Diphosphoglycerate and Phosphoenolpyruvate (PEP) eventually end up?

Figure 6-3: How is it possible to produce a four-carbon butanol by fermenting three-carbon pyruvic acids?

Figure 6-4: How does the number of $CO_2$ molecules produced during respiration compare to the number of carbons in glucose?

What is the name of the process that produces ATP in the electron transport system?

What is the name of the process that produces ATP in glycolysis and the Krebs cycle?

Figure 6-5: What is the source of the Oxaloacetic acid that combines with Acetyl CoA to begin the Krebs cycle?

Succinic acid and Malic acid both have four carbons. In what way do they differ?

Malic acid and Oxaloacetic acid both have four carbons. In what *ways* do they differ?

Which has more potential energy: Succinic acid or Oxaloacetic acid?

Figure 6-7: Notice that the ATP yield of NADH from glycolysis is shown as 4-6 ATP. Explain why the yield varies.

## Matching Questions

Write the letter of the phrase that best matches the numbered term on the left. Use each only once.

_____ 1. cyanobacteria

_____ 2. anaerobe

_____ 3. aerobe

_____ 4. mitochondrion

_____ 5. ATP synthase

_____ 6. alcoholic fermentation

_____ 7. lactic acid fermentation

_____ 8. substrate-level phosphorylation

_____ 9. Krebs cycle

_____10. aerobic respiration

_____11. FAD

_____12. $NAD^+$

_____13. glycolysis

_____14. acetyl coenzyme A

_____15. chemiosmosis

_____16. 2

_____17. 36

_____18. 38

a. organism that survives in the absence of oxygen

b. organism that requires oxygen

c. number of ATPs produced per glucose during aerobic respiration by prokaryotes

d. the products are ethyl alcohol and $CO_2$

e. ATP production associated directly with a carbon compound

f. a coenzyme used in glycolysis and Krebs cycle

g. prokaryotes that evolved $O_2$ producing photosynthesis

h. the enzymes for this are in the mitochondria

i. number of ATPs produced per glucose during aerobic respiration by eukaryotes

j. oxygen is a reactant in the final step

k. mechanism of ATP production in mitochondria

l. a coenzyme used in the Krebs cycle, but not in glycolysis

m. carbon compound that enters the Krebs cycle

n. site of aerobic respiration

o. the conversion of glucose to two pyruvic acids

p. enzyme responsible for ATP synthesis associated with electron transport chain

q. pathway used by exercising muscles to oxidize NADH

r. number of ATPs produced per glucose during fermentation (net)

## Multiple Choice Questions

1. Which DOES NOT describe oxygen?
   a. It is necessary for the survival of anaerobes.
   b. It became abundant in the atmosphere between 2 and 3 billion years ago.
   c. It is destructive to organic chemicals.
   d. Its presence provides an opportunity to extract more energy from organic molecules.
   e. It is not required by all organisms.

2. Which DOES NOT describe aerobes?
   a. They require oxygen.
   b. They have protective mechanisms against the destructive effects of oxygen.
   c. They were among the earliest forms of life to evolve.
   d. They use oxygen to release energy from food molecules.
   e. They include animals as well as plants.

3. Which DOES NOT describe glycolysis?
   a. It occurs in the cytoplasm.
   b. It begins with a phosphorylation of glucose.
   c. It produces $CO_2$ as a product.
   d. It is used by organisms that respire.
   e. It is used by organisms that ferment.

4. Which DOES NOT describe glycolysis?
   a. It is used to make ATP by substrate-level phosphorylation.
   b. Two three-carbon compounds are produced from each glucose.
   c. It requires $O_2$.
   d. It is used by prokaryotic cells.
   e. It is used by eukaryotic cells.

5. Which is a reactant in glycolysis?
   a. ADP
   b. $CO_2$
   c. $H_2O$
   d. pyruvic acid
   e. $O_2$

6. Which is a reactant in glycolysis?
   a. $NADP^+$
   b. $NAD^+$
   c. RuBP
   d. PGAL
   e. NADPH

7. Which is a (net) product of glycolysis?
   a. $O_2$
   b. ADP
   c. glucose
   d. $NADP^+$
   e. pyruvic acid

8. Which is a (net) product of glycolysis?
   a. $CO_2$
   b. NADH
   c. PGAL
   d. RuBP
   e. $H_2O$

9. Which describes fermentation?
   a. All fermenters produce the same end products.
   b. It requires oxygen.
   c. It produces two ATP per glucose.
   d. It is more efficient than aerobic respiration.
   e. No fermenting organisms have survived in the aerobic environment.

10. Which describes fermentation?
    a. All fermenters oxidize NADH.
    b. Most of the energy in glucose is transferred to bonds of ATP during fermentation.
    c. All ATP produced during fermentation comes from reactions involving oxidation of pyruvic acid.
    d. Ethyl alcohol is always a product of fermentation.
    e. The majority of species currently alive are only capable of fermentation.

11. Which DOES NOT describe aerobic respiration?
    a. An electron transport system is involved.
    b. The reactions occur in the cytoplasmic fluid.
    c. It transfers more energy from glucose to the bonds of ATP than fermentation.
    d. It uses a cyclic metabolic pathway.
    e. Coenzyme A is used to make acetyl CoA.

12. Which DOES NOT describe aerobic respiration?
    a. Some ATP is produced by substrate phosphorylation.
    b. The Krebs cycle occurs in mitochondria.
    c. Pyruvic acid from glycolysis is completely oxidized in the reactions of aerobic respiration.
    d. An organic molecule acts as the final electron acceptor.
    e. The Krebs cycle must be run twice in order to completely oxidize glucose.

13. Which is a reactant in the Krebs cycle?
    a. $CO_2$
    b. acetyl CoA
    c. NADPH
    d. $FADH_2$
    e. glucose

14. Which is a reactant in the Krebs cycle?
    a. $O_2$
    b. ATP
    c. $NAD^+$
    d. pyruvic acid
    e. lactic acid

15. Which is a product of the Krebs cycle?
    a. $FADH_2$
    b. ADP
    c. phosphoenol pyruvic acid
    d. acetyl CoA
    e. $NAD^+$

16. Which is a product of the Krebs cycle?
    a. $NADP^+$
    b. $O_2$
    c. pyruvic acid
    d. lactic acid
    e. $CO_2$

17. Which is a reactant in the mitochondrial electron transport system?
    a. pyruvic acid
    b. NADH
    c. $H_2O$
    d. acetyl CoA
    e. lactic acid

18. Which is a reactant in the mitochondrial electron transport system?
    a. $O_2$
    b. $FADH_2$
    c. $CO_2$
    d. $O_2$ and $FADH_2$
    e. $O_2$, $FADH_2$ and $CO_2$

19. Which is a product of the mitochondrial electron transport system?
    a. pyruvic acid
    b. NADH
    c. $H_2O$
    d. acetyl CoA
    e. lactic acid

20. Which is a product of the mitochondrial electron transport system?
    a. $O_2$
    b. $FADH_2$
    c. $CO_2$
    d. FAD
    e. $FADH_2$ and $CO_2$

21. How many ATP can be produced per NADH oxidized in the mitochondrial electron transport system with oxygen as the final electron acceptor?
    a. 1
    b. 2
    c. 3
    d. 24
    e. 36

22. How many ATP can be produced per $FADH_2$ oxidized in the mitochondrial electron transport system with oxygen as the final electron acceptor?
    a. 1
    b. 2
    c. 3
    d. 24
    e. 36

23. Which can be synthesized from intermediates of glycolysis or the Krebs cycle?
    a. carbohydrates
    b. fatty acids
    c. amino acids
    d. nucleotides
    e. All of these can be synthesized from intermediates of glycolysis or the Krebs cycle.

24. Which can be oxidized in glycolysis or the Krebs cycle?
    a. carbohydrates
    b. fatty acids
    c. amino acids
    d. nucleotides
    e. All of these can be oxidized in glycolysis or the Krebs cycle.

## Concept Map Construction

Construct a concept map for each group of terms. Be sure to include appropriate connector phrases. You may add other terms as necessary and use terms in the singular or plural form.

1. NADH, active site, proton gradient, protein, oxidation
2. carbohydrate, mitochondrion, photosystem I, anabolism, cilium
3. homeostasis, aerobic respiration, substrate-level phosphorylation, fermentation, coenzyme

# Chapter 7

# Perpetuating Life: Cell Division

## Go Figure!

Figure 7-1: Does the size of each colored portion for a particular phase represent the length of time a "typical" cell spends in that stage?

Is the sequence from $G_1$ to S to $G_2$ or $G_2$ to S to $G_1$?

According to the diagram, in which specific phase does cytokinesis begin? In which does it end?

Figure 7-2: In which stage do chromosomes appear thickest? They appear thicker at this stage than during the next one. What has made this change?

In which stage do chromosomes appear to be in confused disarray like a plate of spaghetti pasta?

At which stage are chromosomes "pointing" at each other like fingers from across the cell?

At which stage are there two indistinct clumps of chromosomes?

Figure 7-7: List features visible in the photographs you could use to determine which chromosomes are homologous.

What is the significance of bent chromosomes? (Examine pairs 1, 2 and 3 for examples.)

Other than size, how does chromosome 11 differ from chromosome 14?

All of the chromosome pairs are homologous except the X and Y pair shown in the lower right. Why are they grouped if they aren't homologous?

Figure 7-8: One pair of homologous chromosomes is shown in blue. What is the significance of one member of the pair being shown in dark blue and the other in light blue?

What is separated during Anaphase I?

What is separated during Anaphase II?

Is DNA duplicated during the interphase preceding Meiosis II? How can you tell?

Figure 7-9: The shading used in this diagram differs from that used in Figure 7-8. What is the significance of shading one chromosome of the homologous pair blue and the other red?

Assume these are chromosomes in the cells of a new multicellular animal. If the blue chromosome came from the father, where did the red chromosome come from?

Figure 7-10: It usually is not possible to look at chromosomes under the microscope and tell which gene forms are physically located on a chromosome. Since this is the case, how would you be able to tell if a plant chromosome had the gene for tall and the gene for white flowers on the same chromosome?

Figure 7-11: What is the significance of shading one chromosome of a homologous pair green and the other yellow?

What is the significance of having nonhomologous chromosomes shaded the same color?

Is the cell an animal or a plant cell? How can you tell?

What determines which arrangement of green and yellow chromosomes a cell will have? Can you predict the outcome with certainty?

Figure 7-12: Look at the nucleus third down from the top on the left. What are the lines running from right to left? In what visible way are the daughter cells similar?

Compare mitosis and meiosis by looking at the four nuclei shown fourth down from the top. How are the mitotic and meiotic nuclei different? How are they similar?

Continue examining the four nuclei shown fourth down from the top. How are the mitotic and meiotic chromosomes different? How are they similar?

Now compare mitosis and meiosis by looking at the mitotic nuclei fourth down from the top and the meiotic nuclei on the bottom. How are the mitotic and meiotic nuclei different? How are they similar?

Continue examining the mitotic nuclei fourth down from the top and the meiotic nuclei on the bottom. How are the mitotic and meiotic chromosomes different? How are they similar?

## Matching Questions

Write the letter of the phrase that best matches the numbered term on the left. Use each only once.

____ 1. chromatid

____ 2. centromere

____ 3. M phase

____ 4. S phase

____ 5. spindle apparatus

____ 6. metaphase

____ 7. cell plate

____ 8. cleavage furrow

____ 9. zygote

____10. crossing over

____11. maternal chromosome

____12. paternal chromosome

____13. prophase I

____14. prophase II

____15. anaphase I

____16. anaphase II

____17. haploid

____18. gamete

____19. segregation

a. part of cell cycle in which division occurs

b. a strand of densely compacted DNA and protein attached at a centromere to another one like itself

c. made of microtubules

d. formed by fusion of secretion vesicle membranes and their polysaccharide contents

e. time of eukaryotic DNA synthesis

f. typical of animal cytokinesis

g. chromosomes align midway between spindle poles during this phase

h. chromosomal spindle fiber attachment site

i. typically haploid cell

j. cell formed at fertilization

k. an exchange of parts between homologous chromosomes

l. the member of a homologous pair contributed by a sperm

m. meiotic phase when crossing over occurs

n. a member of a homologous pair contributed by the egg

o. phase characterized by separation of homologues

p. applies to a cell which contains one of each type of chromosome present

q. characterized by chromatid separation

r. condensing chromosomes are not paired in this phase

s. occurs during anaphase I

## Multiple Choice Questions

1. Cells which have one of each kind of chromosome are called
   a. prokaryotic.
   b. haploid.
   c. diploid.
   d. cytokinetic.
   e. mitotic.

2. Cells which have two of each kind of chromosome are called
   a. prokaryotic.
   b. haploid.
   c. diploid.
   d. cytokinetic.
   e. mitotic.

3. Which statement DOES NOT describe mitosis?
   a. It is a type of division that produces identical cells.
   b. It is a type of division that may produce generation after generation of identical cells.
   c. In some multicellular organisms, it allows for asexual reproduction.
   d. It is a process which duplicates the hereditary material.
   e. In multicellular organisms, it is responsible for growth.

4. Which statement DOES NOT describe mitosis?
   a. It produces cells with a single molecule of DNA in each chromosome.
   b. It produces two cells with an equal number of chromosomes.
   c. It produces cells with chromosomes composed of two chromatids.
   d. In multicellular organisms, it is responsible for cell replacement.
   e. Chromosomal spindle fibers separate chromatids.

5. Which describes meiosis?
   a. Diploid mother cells produce diploid daughter cells.
   b. Haploid mother cells produce haploid daughter cells.
   c. Two daughter cells are produced from one mother cell.
   d. Diploid mother cells produce haploid daughter cells.
   e. Cytokinesis seldom occurs during or after meiosis.

6. Which describes meiosis?
   a. Haploid mother cells produce diploid daughter cells.
   b. Four daughter cells are produced from one mother cell.
   c. It produces cells with a chromosome number equal to the mother cell.
   d. It produces cells with one half the chromosome number of the mother cell, and each chromosome is composed of two chromatids.
   e. In some multicellular organisms, meiosis produces zygotes.

7. A diploid cell with 24 chromosomes undergoes mitosis. It produces daughter cells with
   a. 12 chromosomes each containing two chromatids.
   b. 12 chromosomes each containing a single molecule of DNA.
   c. 48 chromosomes each containing a single molecule of DNA.
   d. 24 chromosomes each containing two chromatids.
   e. 24 chromosomes each containing a single molecule of DNA.

8. A haploid cell with 24 chromosomes undergoes mitosis. It produces daughter cells with
   a. 12 chromosomes each containing two chromatids.
   b. 12 chromosomes each containing a single molecule of DNA.
   c. 48 chromosomes each containing a single molecule of DNA.
   d. 24 chromosomes each containing two chromatids.
   e. 24 chromosomes each containing a single molecule of DNA.

9. A diploid cell with 24 chromosomes undergoes meiosis. It produces daughter cells with
   a. 12 chromosomes each containing two chromatids.
   b. 12 chromosomes each containing a single molecule of DNA.
   c. 48 chromosomes each containing a single molecule of DNA.
   d. 24 chromosomes each containing two chromatids.
   e. 24 chromosomes each containing a single molecule of DNA.

10. A diploid cell with 36 chromosomes undergoes meiosis. It produces daughter cells with:
    a. 18 chromosomes each containing two chromatids.
    b. 18 chromosomes each containing a single molecule of DNA.
    c. 72 chromosomes each containing a single molecule of DNA.
    d. 36 chromosomes each containing two chromatids.
    e. 36 chromosomes each containing a single molecule of DNA.

11. All the cells of a multicellular plant have been observed to contain 27 chromosomes. This plant is composed of
    a. prokaryotic, diploid cells.
    b. prokaryotic, haploid cells.
    c. eukaryotic, haploid cells.
    d. eukaryotic, diploid cells.
    e. meiotic daughter cells prior to undergoing mitosis.

12. Gametes in a particular animal contain 18 chromosomes. Which is true?
    a. The haploid number is N = 9.
    b. The haploid number is N = 36.
    c. The diploid number is 2N = 18.
    d. The diploid number is 2N = 36.
    e. The diploid number is 2N = 72.

13. Most persons have a total of 46 chromosomes in each body cell. Which is true of these chromosomes?
    a. Twenty-three came from the father and 23 from the mother.
    b. Twenty-two came from the father and 24 from the mother.
    c. All 46 came from the mother.
    d. All 46 came from the father.
    e. The number contributed by each parent is highly variable.

14. Most persons have a total of 46 chromosomes in each body cell. Which is true of these chromosomes?
    a. This is the human haploid number of chromosomes.
    b. This is the human diploid number of chromosomes.
    c. Fertilization will restore the number to 92.
    d. Twenty-three came from the paternal grandfather and 23 from the maternal grandfather.
    e. Twenty-two came from the paternal grandfather and 24 from the maternal grandmother.

15. In which phase does chromosome condensation occur?
    a. interphase
    b. prophase
    c. metaphase
    d. anaphase
    e. telophase

16. In which phase does chromosomal spindle fiber attachment occur?
    a. interphase
    b. prophase
    c. metaphase
    d. anaphase
    e. telophase

17. What event marks the beginning of anaphase?
    a. condensation of chromosomes
    b. alignment of chromosomes at the metaphase plate
    c. synchronous separation of sister chromatids
    d. chromosomes reach their respective spindle poles
    e. chromosomes uncoil

18. What event marks the end of anaphase?
    a. condensation of chromosomes
    b. alignment of chromosomes at the metaphase plate
    c. synchronous separation of sister chromatids
    d. chromosomes reach their respective spindle poles
    e. chromosomes uncoil

19. Cytokinesis in animals results from
    a. contraction of a ring of microtubules.
    b. construction of a new cell membrane and cell wall.
    c. contraction of a ring of microfilaments
    d. construction of a new nuclear membrane.
    e. construction of a cell wall around each new nucleus.

20. Cytokinesis in plants results from
    a. contraction of a ring of microtubules.
    b. construction of a new cell membrane and cell wall.
    c. contraction of a ring of microfilaments
    d. construction of a new nuclear membrane.
    e. construction of a cell wall around each new nucleus.

21. Meiosis increases genetic variability. Some of this variability is produced in a zygote by
    a. mixing paternal and maternal chromosomes during fertilization.
    b. reducing the number of chromosomes after fertilization.
    c. increasing the number of chromosomes after fertilization.
    d. normal mitotic divisions.
    e. the contribution of a haploid number of chromosomes from the sperm and a diploid number by the egg.

22. Meiosis increases genetic variability. Some of this variability results from
    a. mixing paternal and maternal chromosomes during gamete formation.
    b. reducing the number of homologous pairs during meiosis.
    c. increasing the number of homologous pairs during meiosis.
    d. crossing over.
    e. splitting centromeres during anaphase I.

23. During prophase I of meiosis,
    a. the chromosomes unwind to reform chromatin material.
    b. the chromosomes align on the metaphase plate.
    c. crossing over occurs.
    d. centromeres split.
    e. homologous chromosomes separate.

24. During prophase I of meiosis,
    a. tetrads separate.
    b. chromosomes condense.
    c. centromeres form.
    d. cytokinesis occurs.
    e. homologous chromosomes align on the metaphase plate.

Use the following information for questions 25 and 26.

Tomato plants have a gene for fruit shape and another gene for shape of the flower shoot on the same chromosome. The fruit shape gene comes in two forms: one produces round fruit, the other long fruit. The gene for the shape of the flower shoot also comes in two forms: branched and unbranched. For problems 25 and 26, assume that in a diploid plant, one homologue has the gene for round fruit and the gene for unbranched flower shoots. The homologous chromosome has the gene for long fruit and the gene for branched flower shoots.

25. If crossing over DOES NOT occur in tomatoes, what gene combination will be seen in the gametes?
    a. The round fruit gene with the branched flower shoot gene is in all gametes.
    b. Some gametes contain the round fruit gene with the branched flower shoot gene, and other gametes contain the long fruit gene with the branched flower shoot gene.
    c. Some gametes contain the round fruit gene with the unbranched flower shoot gene, and others contain the long fruit gene with the branched flower shoot gene.
    d. Some gametes contain the round fruit gene with the unbranched flower shoot gene, and others contain the long fruit gene with the unbranched flower shoot gene.
    e. Some contain the round fruit gene with the branched flower shoot gene. and others contain the long fruit gene with the unbranched flower shoot gene.

26. If crossing over DOES occur in tomatoes, what NEW gene combinations may be seen in the gametes?
    a. The round fruit gene with the branched flower shoot gene is in all gametes.
    b. Some gametes contain the round fruit gene with the branched flower shoot gene, and other gametes contain the long fruit gene with the branched flower shoot gene.
    c. Some gametes contain the round fruit gene with the unbranched flower shoot gene, and others contain the long fruit gene with the branched flower shoot gene.
    d. Some gametes contain the round fruit gene with the unbranched flower shoot gene, and others contain the long fruit gene with the unbranched flower shoot gene.
    e. Some contain the round fruit gene with the branched flower shoot gene and others contain the long fruit gene with the unbranched flower shoot gene.

27. During anaphase II
    a. the chromosomes unwind to reform chromatin material.
    b. the chromosomes align on the equator of the cell.
    c. the nucleolus is dispersed.
    d. centromeres split.
    e. homologous chromosomes separate.

28. During anaphase II
    a. the chromosomes begin to condense.
    b. synapsis occurs.
    c. chromosomal spindle fibers shorten.
    d. pairs of chromosomes separate.
    e. cytokinesis occurs.

29. In what way do mitosis and meiosis <u>differ</u>?
    a. Mitosis produces cells which are identical; meiosis produces cells which are not identical.
    b. Replication of DNA precedes mitosis; it does not precede meiosis.
    c. A spindle apparatus forms during meiosis; one does not form during mitosis.
    d. Centrioles form during mitosis; they do not form during meiosis.
    e. Chromosomal spindles shorten during mitosis; they do not during meiosis.

30. In what way do mitosis and meiosis <u>differ</u>?
    a. Mitosis results in four new cells; meiosis results in two new cells.
    b. Mitosis produces diploid cells from haploid ones; meiosis produces haploid cells from diploid ones.
    c. Cytokinesis generally occurs once during or after mitosis; it usually occurs at two different points during or after meiosis.
    d. Chromosomal spindles do not shorten during mitosis; they shorten during meiosis.
    e. Centromeres split during mitosis; they do not during meiosis.

31. In what way are mitosis and meiosis similar?
    a. Both produce two new cells from one mother cell.
    b. Both produce haploid cells from diploid cells.
    c. Homologous chromosomes separate during both processes.
    d. Both result in the division of the cytoplasm into two parts.
    e. Chromosomes are moved by a spindle apparatus in both processes.

32. In what way are mitosis and meiosis similar?
    a. Both produce four new cells from one mother cell.
    b. Both produce diploid cells from diploid cells.
    c. Both processes occur only in plants.
    d. Both processes occur in bacteria.
    e. Chromatids separate at some point during both processes.

## Concept Map Construction

Construct a concept map for each group of terms. Be sure to include appropriate connector phrases. You may add other terms as necessary and use terms in the plural or singular form.

1. mitosis, DNA, daughter cell, nucleus, cell plate
2. sexual reproduction, meiosis, haploid, cytokinesis, homologous chromosome
3. mitosis, meiosis, anaphase I, independent assortment, homologue

# Chapter 8

# Fundamentals of Inheritance

## Go Figure!

Figure 8-1: What is the significance of yellow and green peas in the same pod?

What is the purpose of removing the anthers from plants before they have produced pollen?

Why would "accidental" matings have been undesirable during Mendel's experiments?

Why would "fatherless" offspring have complicated Mendel's interpretation of his experiments?

Figure 8-2: In the P generation, one plant was true breeding for round seeds and one for wrinkled seeds.

Describe the physical steps you think Mendel used to cross breed these two plants (refer to Figure 8-1). Do you think he only used pollen from the round seed producing plant? Why or why not?

In the $F_2$ generation, what combinations of peas do you think Mendel saw in a pod?...all round seeds?...all wrinkled seeds?...wrinkled and round mixed?

Figure 8-3: What is the significance of shading one chromosome of a pair blue and the other orange?

How would the ratios be affected by crossing over?

Are both *Aa* offspring sets identical (*i.e.*, *A* and *a* each from the same parent)?

Figure 8-4: Would the probability ratios be changed if male gametes were listed along the side and female gametes along the top?

The figure indicates there is a ratio of 3 round peas to 1 wrinkled pea produced from this cross. What is the relationship between phenotype ratio and the probability of producing a round seed?

Figure 8-6: If a testcross produces 10 peas, all of which are round, what would you conclude about the plant of unknown genotype? Is it possible that the plant of unknown genotype is heterozygous?

What could you do to increase the chance of detecting the presence of a recessive allele in the plant of unknown genotype?

In another testcross, 150 peas are produced, all of which are round. Can you say with certainty that the plant of unknown genotype is homozygous dominant?

Figure 8-7: According to the figure, what is the probability an $F_1$ produced gamete will contain the alleles *RY*? What is the probability it will contain *ry*?

What is the probability that *RY* and *ry* containing gametes will fuse (be fertilized) to form a *RrYy* pea?

What process insures that gametes each contain a single "R" and a single "Y?"

Figure 8-9: What is the probability the male parent will produce a gamete with the genotype ABC?

What is the probability that both parents will produce gametes with the genotype ABC?

What is the probability these parents will produce an *AaBbCc* child?

What is the probability these parents will produce a child intermediate (a 3 on the bottom scale) in skin pigmentation?

Figure 8-11: Observe the larger homologous pair. What is the significance of one allele being shaded red and the other green? Why did the artist draw this homologous pair the same size and shape?

Turn your attention to the smaller pair. What is the significance of both alleles being shaded the same color?

Figure 8-12: This figure illustrates all parental gene combinations. List all nonparental gene combinations.

Figure 8-13: What are the parental gene combinations shown at the top of "b"?

Which gene is involved in cross over in "b"?

What are the nonparental gene combinations which result from crossing over?

The chromosomes in the circles at the bottom of "b" represent all possible cross over and non-cross over combinations. In order for the genes shown to be expressed in the phenotype of an adult fly, what alleles of each gene must the second chromosome contain?

Figure 8-14: According to this Punnett square, what is the probability of having a girl?

What is the probability of having a boy?

Figure 8-16: Which generation is skipped by color blindness in this illustration?

What is another word for carrier?

Why are only females carriers in this figure?

Can a son receive the color blindness allele from his father?

Can a son receive the color blindness allele from a mother with normal color vision?

What phenotypes may the parents have if they have a color blind daughter?

## Matching Questions

Write the letter of the phrase that best matches the numbered term on the left. Use each only once.

____ 1. testcross
____ 2. dominant characteristic
____ 3. recessive characteristic
____ 4. homozygous
____ 5. phenotype
____ 6. Law of Segregation
____ 7. dihybrid cross
____ 8. incomplete dominance
____ 9. codominance
____ 10. pleiotropy
____ 11. mutation
____ 12. wild type
____ 13. linkage group
____ 14. crossing over
____ 15. mutagenic
____ 16. sex chromosomes
____ 17. autosomes
____ 18. Y chromosome
____ 19. carrier
____ 20. oncogene

a. the parental characteristic <u>not</u> shown by a hybrid

b. nonhomologous pair in human male

c. predicts alleles of the same gene will separate prior to gamete formation

d. mating an organism of unknown genotype to one homozygous recessive for the same characteristic

e. cancer causing gene

f. the parental characteristic shown by a hybrid

g. mating parents that differ in two characteristics

h. the physical appearance

i. hybrid expresses both alleles fully

j. heterozygous for a recessive trait

k. a permanent change in a gene

l. a gene affecting many characteristics

m. produces a hybrid phenotypically intermediate to the parents

n. possessing two of the same alleles for a gene

o. unpaired, gender determining in human males

p. genes on the same chromosome

q. describes any agent which causes inheritable changes in DNA

r. explains incomplete linkage

s. homologous pairs in human male

t. the original allele form in a population

## Multiple Choice Questions

1. A gene exists in two different forms: *A* and *a*. An organism with the genotype *Aa* will produce equal numbers of *A*-containing and *a*-containing gametes. This demonstrates
   a. the Law of Segregation.
   b. the Law of Independent Assortment.
   c. recessiveness.
   d. dominance.
   e. homozygosity.

2. Gene "A" exists in two forms: *A* and *a*. Gene "B", located on a different chromosome, also exists in two forms: *B* and *b*. An organism with the genotype *AaBb* will produce equal numbers of *AB*-, *Ab*-, *aB*-, and *ab*-containing gametes. The fate of the "A" and "B" genes relative to each other demonstrates
   a. the Law of Segregation.
   b. the Law of Independent Assortment.
   c. recessiveness.
   d. dominance.
   e. homozygosity.

3. Gene "A" exists in two forms: *A* and *a*. What are *A* and *a* called?
   a. P
   b. $F_1$
   c. genotypes
   d. alleles
   e. hybrids

4. Gene "A" exists in two forms: *A* and *a*. An individual's genotype is known to be *Aa*. Which applies to this individual?
   a. It is homozygous.
   b. It is heterozygous.
   c. It breeds true.
   d. It has two copies of the same allele.
   e. It has the recessive phenotype.

5. Two pea plants breed true: one for round peas, the other for wrinkled. If pollen from the round pea plant is transferred to the wrinkled pea plant, what would you expect the peas to look like?
   a. round peas in all pods.
   b. round peas in some pods, but there is a chance others will have wrinkled peas.
   c. round peas in one half the pods, and wrinkled in the others.
   d. wrinkled peas in some pods, but there is a chance others will have round peas.
   e. wrinkled peas in all pods.

6. Two pea plants breed true: one for round peas, the other for wrinkled. If pollen from the wrinkled pea plant is transferred to the round pea plant, what would you expect the peas to look like?
   a. round peas in all pods.
   b. round peas in some pods, but there is a chance others will have wrinkled peas.
   c. round peas in one half the pods, and wrinkled in the others.
   d. wrinkled peas in some pods, but there is a chance others will have round peas.
   e. wrinkled peas in all pods.

7. Two pea plants breed true: one for tall, the other for dwarf. The plants are crossed, and all hybrid offspring are tall. Which applies to the tall allele?
   a. It is codominant.
   b. It is dominant.
   c. It is incompletely dominant.
   d. It is recessive.
   e. It is partially recessive.

8. Two pea plants breed true: one for tall, the other for dwarf. The plants are crossed, and all hybrid offspring are tall. Which applies to the allele for dwarf?
   a. It is codominant.
   b. It is dominant.
   c. It is incompletely dominant.
   d. It is recessive.
   e. It is partially recessive.

*NOTE: Answers for questions 9 through 12 are provided in the Appendix. Questions 13 through 16 constitute a similar sequence, but the answers are not given.*

Use the following information for questions 9-12.

There is a gene in summer squash plants that determines fruit color. The dominant allele (*Y*) produces yellow fruit and the recessive (*y*) results in green fruit.

9. A *YY* plant will produce ______ fruit.
   a. yellow
   b. green
   c. yellow and green
   d. white
   e. white, yellow, and green

10. What is the genotype of a plant that produces green fruit?
   a. *YY*
   b. *Yy*
   c. *yy*
   d. *YY* or *yy*
   e. *Yy* or *yy*

11. If you want to determine the genotype of a plant that produced 5 yellow fruit, with which genotype would you cross it?
   a. *YY*
   b. *Yy*
   c. *yy*
   d. *YY* or *yy*
   e. *Yy* or *yy*

12. You performed a test cross on a yellow fruit producing plant and 10 offspring all grew into plants that produced yellow fruit. Which would be the most certain conclusion?
   a. The genotype of the yellow fruit plant is *YY*.
   b. The genotype of the yellow fruit plant is *Yy*.
   c. The genotype of the yellow fruit plant is *yy*.
   d. The genotype of the yellow fruit plant is either *YY* or *Yy*.
   e. The genotype of the yellow fruit plant is either *YY* or *yy*.

Use the following information for questions 13-16.

The gene for pea plant height may be represented with a "T" and the gene for flower color with a "C". Let *T* represent the allele for tall, *t* the allele for dwarf, *C* the allele for purple flowers, and *c* the allele for white flowers.

13. A *TTcc* plant will be
    a. tall with white flowers.
    b. dwarf with white flowers.
    c. tall with purple flowers.
    d. short with purple flowers.
    e. tall with purple and white flowers.

14. What is the genotype of a dwarf, white-flowered plant?
    a. *TTCc*
    b. *TtCc*
    c. *TTcc*
    d. *Ttcc*
    e. *ttcc*

15. If you want to determine the genotype of a tall, purple-flowered plant, with which genotype would you cross it?
    a. *TTCc*
    b. *TtCc*
    c. *TTcc*
    d. *Ttcc*
    e. *ttcc*

16. You performed a test cross on a tall, purple-flowered plant and 10 offspring all grew into tall, purple-flowered plants. Which would be the most certain conclusion?
    a. The genotype of the tall, purple-flowered plant is *TTCC*.
    b. The genotype of the tall, purple-flowered plant is *TTCc*.
    c. The genotype of the tall, purple-flowered plant is *TtCc*.
    d. The genotype of the tall, purple-flowered plant is either *TTCC* or *TTCc*.
    e. The genotype of the tall, purple-flowered plant is either *TTCC*, *TTCc*, or *TtCc*.

Use the following information for questions 17-20.

The gene for pea plant height may be represented with a "T" and the gene for flower color with a "C". Let *T* represent the allele for tall, *t* the allele for dwarf, *C* the allele for purple flowers, and *c* the allele for white flowers.

17. What kind of gametes can a *TTcc* individual produce?
    a. *TT*
    b. *cc*
    c. *Tc*
    d. *Tc*, *TT*, and *cc*
    e. *Tc*, *TC*, *TT*, and *cc*

18. What kind of gametes can a *TtCc* individual produce?
    a. *TT*
    b. *Cc*
    c. *Tc* and *tC*
    d. *Tc*, *TT*, and *cc*
    e. *TC*, *Tc*, *tC*, and *tc*

19. If a cross is made between a plant of genotype *TtCc* and one of type *TTcc*, what is the probability that an offspring will be genotype *TTcc*?
   a. 3/4
   b. 1/2
   c. 1/4
   d. 1/8
   e. 1/16

20. If a cross is made between two plants of genotype *TtCc*, what is the probability that an offspring will be genotype *TTcc*?
   a. 3/4
   b. 1/2
   c. 1/4
   d. 1/8
   e. 1/16

Use the following information for questions 21-24.

There is a gene in summer squash plants that determines fruit color. The dominant allele (*Y*) results in yellow fruit and the recessive (*y*) results in green fruit. A plant with the genotype *Yy* has yellowish green fruit. There is a second gene in which the production of white fruit (*W*) is dominant to the production of non-white fruit (*w*). A plant with the genotype *Ww* produces white fruit.

21. If a *Yy* plant is self-pollinated, the results are 1/4 of the fruit are yellow, 1/2 are yellow-green, and 1/4 are green. Which describes this type of allele relationship?
   a. incomplete dominance
   b. one gene affecting the expression of a gene at a different locus
   c. recessiveness
   d. dominance
   e. codominance

22. A plant with the genotype *WwYY* produces white fruit, and a *wwYY* plant produces yellow fruit. Which describes this type of gene relationship?
   a. incomplete dominance
   b. one gene affecting the expression of a gene at a different locus
   c. recessiveness
   d. dominance
   e. codominance

23. A plant with the genotype *WWyy* produces white fruit, and a *wwyy* plant produces green fruit. What proportion of the offspring from a cross of a *WwYy* plant with a *wwyy* plant would produce green fruit?
   a. 1 (100% or absolute certainty)
   b. 1/2
   c. 1/4
   d. 1/8
   e. 0 (Barring a mutation, it can not happen.)

24. An individual with the genotype *WwYy* produces white fruit. An individual with the genotype *wwYy* produces yellowish green fruit. If a *WwYY* plant is crossed with a *Wwyy* plant, what proportion of their offspring is expected to produce yellow fruit?
   a. 1 (100% or absolute certainty)
   b. 3/4
   c. 1/2
   d. 1/4
   e. 0 (Barring a mutation, it can not happen.)

25. Groups of genes on the same chromosome form
    a. homologous pairs.
    b. sex chromosomes
    c. linkage groups.
    d. carriers.
    e. translocations.

26. During meiosis, groups of genes on the same chromosome demonstrate
    a. the Law of Segregation.
    b. the Law of Independent Assortment.
    c. dominance.
    d. linkage.
    e. heterozygosity.

Use the following information for questions 27 and 28.

The diploid number of chromosomes in the *Aedes* genus of mosquitoes is 6. There are no microscopically or genetically detectable sex chromosomes.

27. How many linkage groups are there in an *Aedes* mosquito?
    a. 12
    b. 6
    c. 3
    d. 2
    e. There are too few chromosomes in *Aedes* to have any linkage groups.

28. How many linkage groups are there in the gametes of an *Aedes* mosquito?
    a. 12
    b. 6
    c. 3
    d. 2
    e. There are too few chromosomes in *Aedes* to have any linkage groups.

29. An exchange of genetic material between homologous chromosomes is called
    a. linkage.
    b. crossing over.
    c. translocation.
    d. duplication.
    e. inversion.

30. An exchange of genetic material between nonhomologous chromosomes is called
    a. linkage.
    b. crossing over.
    c. translocation.
    d. duplication.
    e. inversion.

*NOTE: Answers for questions 31 through 33 are provided in the Appendix. Questions 34 through 36 constitute a similar sequence, but the answers are not given.*

Use the following information for questions 31-33.

Genes "A" and "B" are found on the same chromosome. In one individual, allele *A* and *b* are on one homologue and *a* and *B* on the other.

31. What gametes can this organism produce in terms of "A" and "B" alleles if crossing over does not occur?
    a. *AA* and *BB*
    b. *AB* and *ab*
    c. *Ab*, and *aB*
    d. *AB, ab, Ab,* and *aB*
    e. *aa* and *BB*

32. What gametes can this organism produce in terms of "A" and "B" alleles if crossing over occasionally occurs?
    a. *AA* and *BB*
    b. *AB* and *ab*
    c. *Ab*, and *aB*
    d. *AB, ab, Ab,* and *aB*
    e. *aa* and *BB*

33. Assume crossing over does not occur. What genotypes may be seen in the offspring from a *AAbb* individual crossed with a *AaBb* individual?
    a. *AABB*
    b. *AaBb*
    c. *AAbb* and *AaBb*
    d. *AABB, AABb, AAbb, AaBB, AaBb,* and *aabb*
    e. *AABB, AABb, AAbb, AaBB, AaBb, Aabb, aaBB, aaBb,* and *aabb*

Use the following information for questions 34-36.

Genes "A" and "B" are found on the same chromosome. In one individual, allele *A* and *B* are on one homologue and *a* and *b* on the other.

34. What gametes can this organism produce in terms of "A" and "B" alleles if crossing over does not occur?
    a. *AA* and *BB*
    b. *AB* and *ab*
    c. *Ab* and *aB*
    d. *AB, ab, Ab,* and *aB*
    e. *aa* and *BB*

35. What gametes can this organism produce in terms of "A" and "B" alleles if crossing over occurs?
    a. *AA* and *BB*
    b. *AB* and *ab*
    c. *Ab* and *aB*
    d. *AB, ab, Ab,* and *aB*
    e. *aa* and *BB*

36. Assume crossing over occurs. What genotypes may be seen in the offspring from a *AABB* crossed with an *AaBb*?
    a. *AABB*
    b. *AaBb*
    c. *AABB* and *AaBb*
    d. *AABB, AABb, AaBB,* and *AaBb*
    e. *AABB, AABb, AAbb, AaBB, AaBb, Aabb, aaBB, aaBb,* and *aabb*

37. Gender differences in chromosome appearance result from a pair of chromosomes known as
    a. autosomes.
    b. sex chromosomes.
    c. linkage groups.
    d. carriers.
    e. pedigrees.

38. Chromosome pairs identical in both sexes are known as
    a. autosomes.
    b. sex chromosomes.
    c. linkage groups.
    d. carriers.
    e. pedigrees.

39. Part of the human male phenotype may be determined by a single recessive allele
    a. on an autosome.
    b. on the X chromosome.
    c. in a linkage group.
    d. in a translocated portion of an autosome.
    e. in a carrier.

40. Part of the human male phenotype may be determined by a single recessive allele
    a. on an autosome.
    b. in an inversion.
    c. in a duplication.
    d. on the Y chromosome.
    e. in his germ cells.

41. Chromosomal aberrations include a change in the sequence of genes on a chromosome. This is known as
    a. a carrier.
    b. an inversion.
    c. a duplication.
    d. a translocation.
    e. polyploidy.

42. Chromosomal aberrations include a change in the number of sets of chromosomes. The result may change cells from diploid (2n) to triploid (3n). This is known as
    a. a carrier.
    b. an inversion.
    c. a duplication.
    d. a translocation.
    e. polyploidy.

## Concept Map Construction

Construct a concept map for each group of terms. Be sure to include appropriate connector phrases. You may add other terms as necessary.

1. hybrid, $F_1$, Law of Segregation, dominant, phenotype
2. Law of Independent Assortment, linkage group, X chromosome, gene, inversion
3. homologous chromosome, mutant, gene, crossing over, testcross

# Chapter 9

# The Molecular Basis of Genetics

## Go Figure!

Figure 9-1: To which sugar carbon atom is the phosphate attached?

To which sugar carbon atom is the nitrogenous base attached?

What kind of bond is the sugar-phosphate linkage?

Observe Figure 9-1b. How many nucleotides are shown in this polynucleotide?

Figure 9-2: In what obvious way would the DNA molecule be altered if base pairing involved complementarity between guanine and adenine, and cytosine and thymine? (You may want to look at Figure 9-3 before answering.)

Figure 9-3: Name the bases for each nucleotide in this schematic representation of DNA.

Figure 9-4: Assume this is eukaryotic DNA. After DNA replication, there are four strands of DNA: two original and two newly synthesized. How are these four strands distributed in the chromatids of a eukaryotic chromosome?

Figure 9-5: Synthesis of a new DNA strand occurs in which direction: 5' to 3' or 3' to 5'?

Why must one strand always be synthesized discontinuously?

Figure 9-7: Since only one strand of DNA (the "sense strand") is necessary for transcription, why is DNA double stranded?

If it had not been indicated for you in this figure, how could you have deduced the direction of transcription?

Figure 9-8: What does the ability of tRNA to base pair with itself tell you about the linear sequence of nucleotides within the molecule?

Figure 9-9: How many different amino acids are shown in the chart?

Which amino acids are encoded by the most codons?

Which amino acids are encoded by the fewest codons?

Which nucleotide (first, second or third) in a codon appears to be least important in coding for an amino acid?

Figure 9-12: Does the figure illustrate transcription of more than one gene? Justify your answer.

Figure 9-13: What is accomplished by packaging eukaryotic DNA?

Suggest reasons why this degree of packaging is not seen in prokaryotes.

## Matching Questions

Write the letter of the phrase that best matches the numbered term on the left. Use each only once.

____ 1. adenine
____ 2. cytosine
____ 3. uracil
____ 4. tRNA
____ 5. rRNA
____ 6. mRNA
____ 7. replication fork
____ 8. semiconservative replication
____ 9. nucleosome
____10. histone
____11. frameshift mutation
____12. mutagen
____13. transcription
____14. translation
____15. complementarity
____16. DNA polymerase
____17. RNA polymerase
____18. codon
____19. anticodon
____20. nonsense codon

a. three of its nucleotides make up the anticodon
b. A always pairs with T and G always pairs with C
c. a unit of DNA packaging
d. RNA synthesis
e. occurs as a result of insertion or deletion of a nucleotide
f. the enzyme of DNA replication
g. three adjacent nucleotides on tRNA
h. a purine of DNA
i. three nucleotides on mRNA for which there is no anticodon
j. the site of DNA strand separation
k. a structural component of ribosomes
l. three adjacent nucleotides on mRNA
m. the enzyme of transcription
n. protein associated with DNA
o. a pyrimidine of DNA
p. protein synthesis
q. both strands act as a template during DNA synthesis
r. it contains the codons
s. physical agent that changes DNA
t. the nucleotide that replaces thymine in RNA

## Multiple Choice Questions

1. Who was responsible for identifying DNA as the molecule responsible for transformation in bacteria?
   a. Franklin
   b. Avery
   c. Watson and Crick
   d. Hershey and Chase
   e. Griffith

2. Who is credited with determining DNA structure?
   a. Franklin
   b. Avery
   c. Hershey and Chase
   d. Watson and Crick
   e. Griffith

3. Which DOES NOT describe DNA?
   a. It forms a double helix.
   b. Adjacent nucleotides are covalently bonded together.
   c. Each nucleotide contains one of four nitrogenous bases.
   d. The sugar in each nucleotide is ribose.
   e. The chains of nucleotides run in "opposite" directions.

4. Which DOES NOT describe DNA?
   a. The nitrogenous bases extend towards the center of the molecule.
   b. Complementary bases are hydrogen bonded to each other.
   c. An adenine nucleotide in one strand base pairs with a thymine nucleotide in the other.
   d. The number of purine nucleotides equals the number of pyrimidine nucleotides in any DNA strand.
   e. The ends of each strand are identified as 3' or 5'.

5. Which describes the prokaryotic chromosome?
   a. It is a circular molecule of DNA.
   b. It is composed of DNA and histone proteins.
   c. When fully extended, it is shorter than the length of the bacterial cell.
   d. It is organized into nucleosomes.
   e. It is located within the nucleus.

6. Which DOES NOT describe the eukaryotic chromosome?
   a. It is made of a linear DNA molecule.
   b. Each chromosome is associated with histone proteins.
   c. Chromosomal DNA is organized into nucleosomes.
   d. Chromosomes are located in the nucleus.
   e. It is shorter than a prokaryotic chromosome.

7. Which DOES NOT describe DNA replication?
   a. DNA polymerase is responsible for inserting new nucleotides.
   b. Replication is semiconservative.
   c. One new strand undergoes continuous synthesis, whereas the other undergoes discontinuous synthesis.
   d. The replication fork is the site of strand separation.
   e. New strand synthesis is in a 3' to 5' direction.

8. Which DOES NOT describe DNA replication?
   a. Each strand acts as a template for synthesis of a new strand.
   b. Synthesis towards the replication fork is continuous.
   c. There are RNA molecules responsible for unwinding the DNA helix.
   d. Mutations may be caused by mistakes during replication.
   e. Prokaryotic DNA polymerase has an active site that allows it to "proofread" the accuracy of nucleotide insertion.

9. What is the complementary strand of DNA to the DNA sequence below?

   3'GTAGCCGTAACGTAT5'

   a. 3'CATCGGCATTGCATA5'
   b. 3'CAUCGGCAUUGCAUA5'
   c. 5'CATCGGCATTGCATA3'
   d. 5'CAUCGGCAUUGCAUA3'
   e. 5'GTAGCCGTAACGTAT3'

10. What is the complementary strand of DNA to the DNA sequence below?

    5'TCGACGGTAGCCTGTG3'

    a. 3'AGCTGCCATCGGACAC5'
    b. 3'AGCUGCCAUCGGACAC5'
    c. 5'AGCTGCCATCGGACAC3'
    d. 5'AGCUGCCAUCGGACAC3'
    e. 5'TCGACGGTAGCCTGTG3'

11. What macromolecule is synthesized during transcription?
    a. DNA
    b. RNA
    c. protein
    d. carbohydrate
    e. lipid

12. What macromolecule is synthesized during translation?
    a. DNA
    b. RNA
    c. protein
    d. carbohydrate
    e. lipid

13. What are the components of an RNA nucleotide?
    a. deoxyribose sugar
    b. deoxyribose sugar and a phosphate
    c. deoxyribose sugar, a phosphate, and a nitrogenous base
    d. ribose sugar and a phosphate
    e. ribose sugar, a phosphate, and a nitrogenous base

14. In what way do RNA nucleotides differ?
    a. They have different sugars.
    b. They have different phosphates.
    c. They have different nitrogenous bases.
    d. They have different bonds holding them together.
    e. There are no differences between RNA nucleotides.

15. Which DOES NOT describe codons?
    a. Each is made of three adjacent nucleotides.
    b. They are part of a tRNA molecule.
    c. They code for amino acid sequence in protein.
    d. Their "meaning" is virtually universal among species.
    e. There are 64 different codons.

16. Which DOES NOT describe anticodons?
    a. Each is made of three adjacent nucleotides.
    b. They are part of a tRNA molecule.
    c. They base pair with nucleotides in rRNA.
    d. Each corresponds to a particular amino acid.
    e. They are an essential component of translation.

17. Which describes a ribosome?
    a. It consists of two subunits.
    b. It is always in functional form.
    c. It contains a codon.
    d. It contains an anticodon.
    e. It is the site of transcription.

18. Which describes a ribosome?
    a. It is made exclusively of rRNA.
    b. The large subunit contains the enzyme that forms the bond between amino acids.
    c The small subunit contains the enzyme that forms the bond between amino acids.
    d. The large subunit binds to the initiation (start) codon (AUG).
    e. The large subunit binds to the initiation (start) anticodon (AUG).

NOTE: *Answers for questions 19 through 22 are provided in the Appendix. Questions 23 through 26 constitute a similar sequence, but the answers are not given.*

19. What is the mRNA transcribed from the stretch of DNA below?

    3'TACGGCACGTAT...5'

    a. 3'AUGCCGUGCAUA...5'
    b. 3'ATGCCGUGCAUA...5'
    c. 5'AUGCCGUGCAUA...3'
    d. 5'ATGCCGUGCAUA...3'
    e. 3'UACGGCACGUAU...5'

20. What is the amino acid sequence coded by the DNA in question #19?
    a. isoleucine-arginine-alanine-valine-...
    b. tyrosine-glycine-threonine-tyrosine-...
    c. methionine-proline-cysteine-isoleucine-...
    d. tyrosine-alanine-arginine-histidine-...
    e. methionine-arginine-serine-isoleucine-...

21. Assume an adenine is inserted between the neighboring guanines of the DNA in question #19. What is the mRNA transcribed from this mutant DNA?
    a. 3'AUGCACGUGCAUA...5'
    b. 3'ATGCACGUGCAUA...5'
    c. 5'AUGCUCGUGCAUA...3'
    d. 5'ATGCACGUGCAUA...3'
    e. 3'UACGUGCACGUAU...5'

22. What is the amino acid sequence coded by the mutant DNA in question #21?
   a. isoleucine-arginine-arginine-serine-...
   b. tyrosine-valine-histidine-isoleucine-...
   c. methionine-proline-cysteine-isoleucine-...
   d. tyrosine-histidine-valine-histidine-...
   e. methionine-leucine-valine-histidine-...

23. What is the mRNA transcribed from the stretch of DNA below?

3'TACGAACTAGCG...5'

   a. 5'AUGCUUGAUCGC...3'
   b. 5'ATGCTUGATCGC...3'
   c. 3'UACGAACUAGCG...5'
   d. 3'AUGCUUGAUCGC...5'
   e. 3'ATGCTTGATGCG...5'

24. What is the amino acid sequence coded by the DNA in question #23?
   a. methionine-leucine-aspartic acid-arginine-...
   b. tyrosine-glutamic acid-leucine-alanine-...
   c. tyrosine-glutamic acid-isoleucine-alanine-...
   d. methionine-glutamic acid-leucine-alanine-...
   e. valine-leucine-aspartic acid-arginine-...

25. Assume the adenine in the fifth position of the DNA in question #23 is deleted. What is the mRNA transcribed from this mutant DNA?
   a. 3'UACGACUAGCG...5'
   b. 3'AUGCUGAUCGC...5'
   c. 3'ATGCUGATGCG...5'
   d. 5'ATGCUGATCGC...3'
   e. 5'AUGCUGAUCGC...3'

26. What is the amino acid sequence coded by the mutant DNA in question #25?
   a. tyrosine-glycine-stop
   b. methionine-glycine-stop
   c. methionine-leucine-isoleucine-alanine-...
   d. methionine-proline-isoleucine-...
   e. tyrosine-glycine-isoleucine-stop

## Concept Map Construction

Construct a concept map for each group of terms. Be sure to include appropriate connector phrases. You may add other terms as necessary and use terms in the singular or plural form.

1. tRNA, ribosome, semiconservative replication, allele, hypothesis testing
2. protein, Golgi complex, histidine, start codon, homologous chromosome
3. crossing over, frameshift mutation, monohybrid cross, DNA polymerase, metaphase

# Chapter 10

# Gene Expression: Orchestrating Life

## Go Figure!

Figure 10-1: Your text says humans and sea squirts belong to the same phylum. On a genetic level, how do you account for the similarities taxonomists used to place them in the same phylum?

On a genetic level, how do you account for their obvious differences?

Figure 10-2: Based on the illustration, suggest ways in which the regulatory protein might be able to inactivate a gene.

Figure 10-3: What is the role of the regulatory gene?

What is the role of the operator?

What is the role of the promoter?

Justify the statement: "Bacterial operons provide a more efficient means of genetic control over metabolic pathways than eukaryotic systems of genetic control."

Figure 10-4: Prokaryotic anabolic pathways often use the final product as a *corepressor*. What effect should an accumulation of this corepressor (end product) have on the operon?

What effect should the corepressor have on the repressor?

Figure 10-6: In what ways are eukaryotic genetic regulation (as shown in this figure) similar to prokaryotic genetic regulation (as shown in Figure 10-4).

In what ways do the two differ?

Figure 10-7: Why do you suppose DNA forms puffs when it is being transcribed?

Figure 10-9: Which part(s) of the eukaryotic gene shown correspond(s) to a prokaryotic gene?

## Matching Questions

Write the letter of the phrase that best matches the numbered term on the left. Use each only once.

____ 1. totipotent

____ 2. clone

____ 3. regulatory gene

____ 4. inducer

____ 5. repressor

____ 6. operator

____ 7. promoter

____ 8. structural gene

____ 9. differentiation

____10. exon

____11. intron

____12. primary transcript

____13. transposition

____14. chromosomal puff

____15. transcription

____16. ecdysone

____17. RNA processing

____18. testosterone receptor protein

a. DNA that codes for an enzyme

b. portion of DNA where repressor binds

c. process of making specialized cells

d. converts primary transcript into mRNA

e. describes a cell able to give rise to a whole new individual

f. a coding part of a eukaryotic gene

g. an insect hormone responsible for metamorphosis

h. movement by "jumping genes"

i. binds to repressor and inactivates it

j. a defective gene for this produces androgen insensitivity syndrome

k. RNA that must be processed to remove introns

l. offspring that is genetically identical to the single parent

m. site of transcription

n. portion of DNA where RNA polymerase binds

o. a noncoding part of a eukaryotic gene

p. codes for the repressor

q. producing RNA from DNA

r. binds to the operator site of operons

## Multiple Choice Questions

1. Which DOES NOT describe gene regulatory proteins?
   a. Some are activated as a result of chemical changes in the external environment.
   b. Some are activated as a result of physical changes in the external environment.
   c. Some are activated as a result of hormonal changes in the internal environment.
   d. Some are activated as a result of conscious decisions of the organism.
   e. They are responsible for determining which genes are used in a particular cell at any given time.

2. Which DOES NOT describe gene regulatory proteins?
   a. They are specific for the genes they regulate.
   b. Some are responsible for activating the gene.
   c. Some are responsible for inactivating the gene.
   d. Their precise mechanism of action was easily determined by molecular biologists.
   e. Some are shaped so as to fit the structure of DNA.

3. Which codes for an enzyme?
   a. operator
   b. repressor
   c. structural gene
   d. regulator gene
   e. intron

4. Which codes for repressor protein?
   a. operator
   b. repressor
   c. structural gene
   d. regulator gene
   e. intron

5. Which describes operons?
   a. They are found in prokaryotes.
   b. They consist only of a regulatory gene.
   c. They may only be switched off (repressed).
   d. They may only be switched on (induced).
   e. They may contain a promoter site for producing repressor protein.

6. Which describes operons?
   a. They are found in eukaryotes.
   b. They contain structural genes for enzymes of a metabolic pathway.
   c. Each structural gene of an operon is regulated independently.
   d. The operator binds RNA polymerase.
   e. The regulatory gene determines whether or not the operon is transcribed.

7. What is the role of the operator in the lac operon?
   a. It binds to RNA polymerase.
   b. It produces the repressor protein.
   c. It produces the inducer.
   d. It binds to the repressor.
   e. It binds to the corepressor.

8. What is the role of the promoter in the lac operon?
   a. It binds to RNA polymerase.
   b. It produces the repressor protein.
   c. It produces the inducer.
   d. It binds to the repressor.
   e. It produces corepressor.

9. What is the role of lactose in the lac operon?
   a. It is a repressor.
   b. It is an inducer.
   c. It is a corepressor.
   d. It is a substrate for RNA polymerase.
   e. It is a substrate for DNA polymerase.

10. What is the role of the repressor protein in the lac operon?
   a. It binds to the promoter.
   b. It binds to the operator.
   c. It binds to RNA polymerase.
   d. It binds to DNA polymerase.
   e. It binds to the corepressor.

11. What happens to an inducible operon when the inducer is present?
   a. The repressor starts being produced.
   b. RNA polymerase binds to the inducer.
   c. The operator is bound to repressor protein.
   d. Transcription occurs.
   e. The corepressor is inhibited.

12. In many anabolic pathways, the final product acts as the effector in regulation of the operon. Which is a logical mechanism of action? The effector binds to
   a. the repressor and inactivates it.
   b. the repressor and activates it.
   c. RNA polymerase and activates it.
   d. DNA polymerase and inactivates it.
   e. the operator and activates it.

13. What is an intron?
   a. It is a region of DNA located between genes that is not translated into protein.
   b. It is a region of DNA located within a gene that is not translated into protein.
   c. It is a regulatory gene.
   d. It is a structural gene.
   e. It is an enzyme.

14. What is an exon?
   a. It is a regulatory gene.
   b. It binds to corepressor.
   c. It is a segment of a split gene that codes for amino acids in a protein.
   d. It is a segment of a split gene that has no corresponding amino acids in a protein.
   e. It is a DNA spacer between genes.

15. Which describes a eukaryotic transcription-level control mechanism?
   a. It affects whether or not mRNA is made.
   b. It regulates an operon.
   c. It affects protein synthesis by interfering with assembly of ribosomal subunits.
   d. It affects gene expression by altering the primary RNA transcript by removing noncoding segments.
   e. It exerts its control in the cytoplasm.

16. Which describes a eukaryotic processing-level control mechanism?
   a. It affects protein synthesis by interfering with the ability of the small ribosomal subunit to bind mRNA.
   b. It affects the frequency of transcription by interfering with RNA polymerase.
   c. It affects the frequency of translation by disrupting codon-anticodon interactions.
   d. It affects the conversion of the primary RNA transcript into functional mRNA that can be translated.
   e. It affects the longevity of mRNA molecules in the cytoplasm.

17. Which describes ecdysone?
    a. It is a hormone that promotes translation of genes encoding larval proteins.
    b. It is a hormone that promotes transcription of genes encoding proteins necessary for metamorphosis in insects.
    c. It is a hormone found only in male insects that promotes development into an adult male.
    d. It is a hormone found only in female insects that promotes development into an adult female.
    e. It is a cytoplasmic receptor protein that combines with a hormone to initiate development of the larval insect into an adult.

18. Which best describes testosterone receptor protein?
    a. The gene for testosterone receptor protein is present in all human cells.
    b. The gene for testosterone receptor protein is present only in some male human cells.
    c. Testosterone receptor protein is bound to membrane.
    d. Testosterone receptor protein is present in all male cells.
    e. Testosterone receptor protein binds to ribosomes to promote translation of the hormone testosterone.

19. Which provides visual evidence of transcription?
    a. testosterone receptor protein
    b. RNA processing
    c. chromosomal puffs
    d. ecdysone
    e. an operon

20. Which regulates gene expression at the translational level?
    a. repressor protein
    b. chromosomal puffs
    c. hormone receptor proteins
    d. RNA polymerase
    e. prolactin

## Concept Map Construction

Construct a concept map for each group of terms. Be sure to include appropriate connector phrases. You may add other terms as necessary and use terms in the singular or plural form.

1. transcriptional-level control, chromosomal puff, mRNA degradation, translation, eukaryote
2. intron, operon, catabolic pathway, DNA, nucleolus
3. interphase, DNA polymerase, microtubule, cell differentiation, homozygote

# Chapter 11

# Genetics and Human Life

## Go Figure!

Figure 11-1: Assume the chromosome involved in nondisjunction is chromosome #21. Which gamete(s) in this figure would produce Down syndrome individuals when combined with a normal gamete at fertilization?

Now assume the chromosome involved in nondisjunction is the X chromosome. Which gamete(s) in this figure would produce an individual with Klinefelter syndrome?

Which gamete(s) in this figure would produce an individual with Turner syndrome?

Figure 11-2: Observe Figure 11-2a. What does bp stand for?

Observe the RFLP patterns in Figure 11-2c. How many different patterns are possible in the children of these parents?

What was the direction of fragment migration in the gel represented?

Figure 11-4: Assume the parents in this figure have three children: one is normal and two are carriers. What is the probability their fourth child will be affected (homozygous recessive)?

Figure 11-5: Assume the parents shown in Figure 11-5a have three children, all of whom are affected (heterozygous). What is the probability the fourth child will be normal?

What are the genotypes of the original parents shown in the pedigree in Figure 11-5b? (It may be impossible to tell with any certainty.)

What are the genotypes of their six children? (In some cases, it may be impossible to tell with any certainty.)

What are the genotypes of their seven grandchildren? (In some cases, it may be impossible to tell with any certainty.)

Figure 11-8: What does the term "transgenic" mean?

Why were littermates used in this experiment?

How was the rat growth hormone gene introduced into all of the mouse cells?

What problems might be encountered if this procedure was attempted with adult mice?

Figure 11-9: Plasmids are used to introduce foreign genes into bacteria. How might plasmid antibiotic resistance genes be useful to the genetic engineer in determining what bacterial cells have taken in the plasmid (and therefore the foreign gene)?

Figure 11-10: The restriction enzyme Eco R1 always cuts DNA between the T and C of a GAATTC sequence. Other restriction enzymes are Hind III and Bam HI. Hind III cuts between the two T's in an AAGCTT sequence. Bam HI cuts between the two Cs of a GGATCC sequence. Write each of these sequences along with their complementary strands. What pattern is there in the nucleotide sequences "recognized" by these restriction enzymes?

Figure 11-11: Since both the plasmid and the insulin gene (or any gene used in this procedure) have the same sticky ends, what prevents the plasmid from joining with itself and not the insulin gene?

As shown in Figure 11-9, some plasmids have more than one gene for antibiotic resistance. Only bacteria that have taken up the plasmid will grow in an environment containing ampicillin since they now have the ampicillin resistance gene. But this doesn't tell the genetic engineer that the plasmid in the bacterial cell has the desired gene (insulin, in this case). How might the second antibiotic resistance gene be used to help the genetic engineer discriminate between those cells that picked up "empty" plasmids and those that picked up recombinant plasmids? (Hint: suppose there is a restriction site within the tetracycline resistance gene.)

Figure 11-13: Why must the sample be alternately heated and cooled in PCR?

What are the advantages of PCR over cloning using bacterial plasmids?

How might the original DNA fragment used in PCR be obtained and identified?

## Matching Questions

Write the letter of the phrase that best matches the numbered term on the left. Use each only once.

____ 1. gene therapy
____ 2. meiotic nondisjunction
____ 3. genetic disorder
____ 4. carrier of a gene
____ 5. electrophoresis
____ 6. autosomal recessive allele
____ 7. X-linked recessive allele
____ 8. amniocentesis
____ 9. RFLPs
____10. recombinant DNA
____11. genetic engineering
____12. gene splicing
____13. vector
____14. plasmid
____15. restriction enzyme
____16. DNA ligase
____17. DNA probe
____18. polymerase chain reaction
____19. DNA fingerprint
____20. transgenic animal

a. defective gene or chromosome
b. technique used to separate restriction fragments
c. a single allele in males produces the phenotype, whereas two are necessary in females
d. carried on nonsex chromosome, but must be homozygous to produce its phenotype
e. replacement of a faulty gene
f. procedure in which fetal cells are isolated and cultured
g. refers to a heterozygote, if the gene is inherited as an autosomal recessive
h. failure of homologs or chromatids to separate
i. laboratory process of creating novel DNA molecules from different biological sources
j. plasmids and viruses are examples
k. process used for enzymatic DNA amplification
l. single stranded DNA used to locate desired restriction fragments after electrophoresis
m. a DNA molecule constructed from pieces derived from different sources
n. enzyme used to covalently join DNA fragments
o. small, circular bacterial DNA molecules
p. gel pattern of unique DNA fragments
q. enzyme that cuts DNA and leaves "sticky ends"
r. insertion of the desired gene DNA into vector DNA
s. produced by restriction enzymes acting on different DNA molecules
t. this organism contains genes from a different species

## Multiple Choice Questions

*NOTE: Answers for questions 1 through 3 are provided in the Appendix. Questions 4 through 6 constitute a similar sequence, but the answers are not given.*

Use the following information for questions 1 through 3.

Tay-Sachs disease is caused by an autosomal recessive allele that results in deficiency of the enzyme hexoaminidase A. Symptoms include blindness and retardation. Onset of symptoms begins at about six months of age, and death results in early childhood. Genetic counseling of at-risk populations (Ashkenazic Jews and French Canadians in Quebec) has reduced the frequency of the disease.

1. What is the probability that a homozygous normal man and a carrier female will have Tay-Sachs child?
   a. 0%
   b. 25%
   c. 50%
   d. 75%
   e. 100%

2. What is the probability that the parents in question #1 will have a child who is a Tay-Sachs carrier?
   a. 0%
   b. 25%
   c. 50%
   d. 75%
   e. 100%

3. Alicia's parents are both normal, but she had a sister who died of Tay-Sachs. What is the probability that Alicia is a carrier of the Tay-Sachs allele?
   a. 0%
   b. 25%
   c. 50%
   d. 75%
   e. 100%

Use the following information for questions 4 through 6.

Ellis-van Creveld syndrome is a type of dwarfism inherited as an autosomal recessive disease. It is rare in the general population of the United States, but it is unusually common in the Amish people of Pennsylvania. People with this disease have a normal body length, but have shortened limbs and appendages as well as extra fingers or toes.

4. What is the probability that two heterozygotes will have a child that suffers from Ellis-van Creveld syndrome?
   a. 0%
   b. 25%
   c. 50%
   d. 75%
   e. 100%

5. What is the probability that an Ellis-van Creveld carrier and a normal homozygous individual will have a carrier child?
   a. 0%
   b. 25%
   c. 50%
   d. 75%
   e. 100%

6. Dick suffers from Ellis-van Creveld syndrome. Margie has a normal stature, but her father suffered from Ellis-van Creveld syndrome. What is the probability that Dick and Margie will produce a dwarf child?
   a. 0%
   b. 25%
   c. 50%
   d. 75%
   e. 100%

*NOTE: Answers for questions 7 through 10 are provided in the Appendix. Questions 11 through 14 constitute a similar sequence, but the answers are not given.*

Use the following information for questions 7 through 10.

Lesch-Nyhan syndrome is produced by a recessive allele carried on the X chromosome. Symptoms include compulsive chewing of the lips, fingers, and toes, as well as other abnormal activities. Since this self-destructive behavior is eventually lethal, no females with Lesch-Nyhan syndrome have been identified.

7. What is the probability of a carrier and a normal male having a carrier child?
   a. 0%
   b. 25%
   c. 50%
   d. 75%
   e. 100%

8. What is the probability the sons in question #7 will have Lesch-Nyhan syndrome?
   a. 0%
   b. 25%
   c. 50%
   d. 75%
   e. 100%

9. Nathan is a normal teenager. However, he had a brother who died of Lesch-Nyhan syndrome. What is the probability that Nathan is a carrier of the allele for this disease?
   a. 0%
   b. 25%
   c. 50%
   d. 75%
   e. 100%

10. Refer to question #9. What are the genotypes of Nathan's parents? ($X^*$ = Lesch-Nyhan allele.)
   a. XY and XX
   b. $X^*Y$ and XX
   c. XY and $X^*X^*$
   d. $X^*Y$ and $X^*X^*$
   e. XY and $X^*X$

Use the following information for questions 11 through 14.

Hemophilia A is a disease in which patients lack a clotting factor in the blood (factor VIII). It is inherited as an X-linked recessive trait.

11. What is the probability of a hemophilic male and a carrier female producing a hemophilic daughter?
   a. 0%
   b. 25%
   c. 50%
   d. 75%
   e. 100%

12. What is the probability that the parents in question #11 will produce a hemophilic son?
   a. 0%
   b. 25%
   c. 50%
   d. 75%
   e. 100%

13. Eric has normal blood. His wife Jane is the daughter of a hemophilic father and a homozygous normal mother. What are the chances that Eric and Jane will have a daughter who is a carrier?
   a. 0%
   b. 25%
   c. 50%
   d. 75%
   e. 100%

14. Ralph has normal blood, but he has two brothers and one sister who are hemophilic. What are the genotypes of Ralph's parents? ($X^H$ = normal; $X^h$ = hemophilia)

a. $X^HY$ and $X^HX^H$
b. $X^hY$ and $X^HX^H$
c. $X^HY$ and $X^hX^h$
d. $X^hY$ and $X^hX^h$
e. $X^hY$ and $X^HX^h$

Questions 15 and 16 use a pedigree to show the inheritance pattern of particular traits in a family. A key to the standard symbols used in constructing pedigrees is provided below.

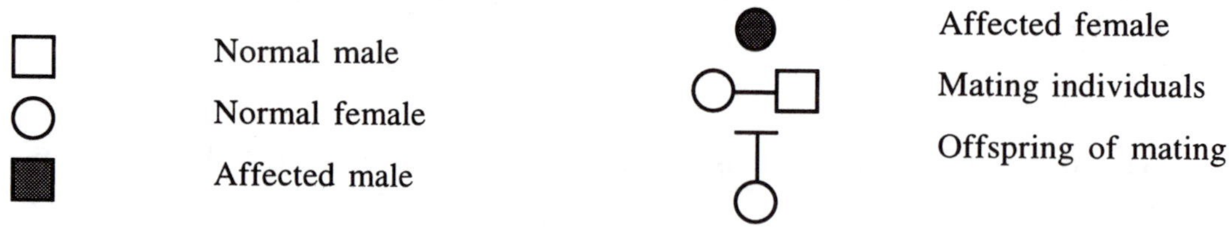


15. Choose the mode of inheritance that best fits the hypothetical pedigree.

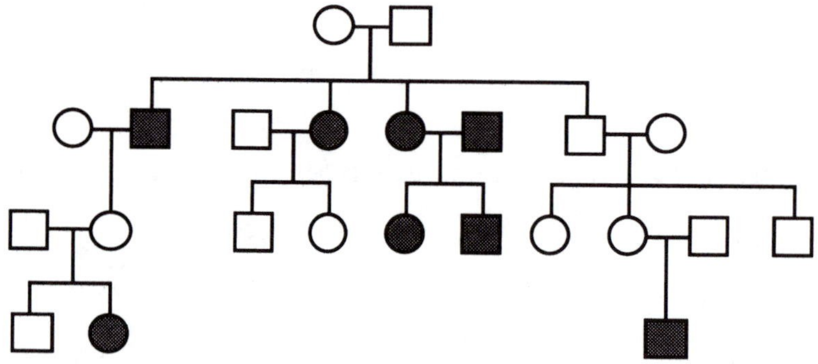

a. The gene is inherited as an X-linked recessive.
b. The gene is inherited as an X-linked dominant.
c. The gene is inherited as an autosomal dominant.
d. The gene is inherited as an autosomal recessive.
e. The gene is inherited as a Y-linked dominant.

16. Choose the mode of inheritance that best fits the hypothetical pedigree.

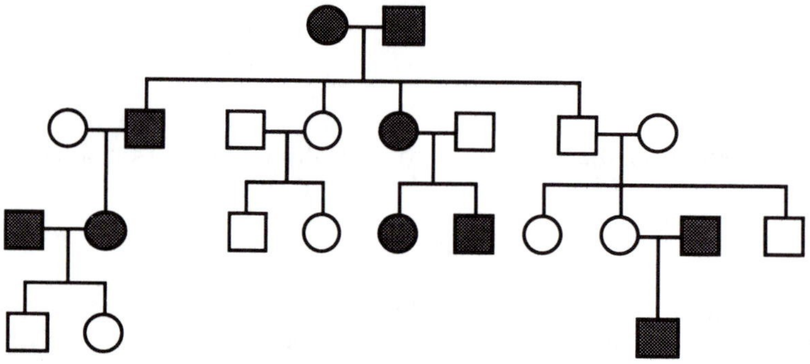

a. The gene is inherited as an X-linked recessive.
b. The gene is inherited as an X-linked dominant.
c. The gene is inherited as an autosomal dominant.
d. The gene is inherited as an autosomal recessive.
e. The gene is inherited as a Y-linked dominant.

17. Which describes genetic engineering?
    a. It occurs frequently in nature.
    b. It occurs in nature, but not frequently.
    c. It involves combining DNA from individuals of the same species.
    d. It involves the production of recombinant DNA molecules.
    e. It requires enzymes artificially produced in the laboratory.

18. Which describes genetic engineering?
    a. It occurs in the laboratory under human direction.
    b. It occurs in the laboratory, but relies on chance to produce new genotypes.
    c. It relies on mating individuals of the same species to produce new genotypes.
    d. It relies on mating individuals of different species to produce new genotypes.
    e. It has only been successful using bacterial DNA.

19. Which is typically used as a vector in genetic engineering?
    a. bacterial chromosome
    b. bacterial plasmid
    c. human chromosome
    d. bacterial RNA
    e. plant chromosome

20. Which is typically used as a vector in genetic engineering?
    a. animal chromosome
    b. viral DNA
    c. viral RNA
    d. viral plasmid
    e. antisense strand

21. Which describes restriction enzymes?
    a. They form covalent bonds between DNA fragments.
    b. They are creations of genetic engineers and do not exist naturally.
    c. Each acts only on specific nucleotide sequences.
    d. They remove introns.
    e. They act on specific codon sequences in mRNA.

22. Which describes restriction enzymes?
    a. They often produce "staggered cuts" in DNA.
    b. Different restriction enzymes are used to cut vector DNA and DNA with the desired gene.
    c. They remove exons from RNA.
    d. They remove exons from DNA.
    e. They make clones of the desired gene.

23. What is the next step in a genetic engineering project after a DNA fragment with a desired human gene has been produced?
    a. Plasmid DNA must be cut with the same ligase used to cut the human DNA.
    b. Plasmid DNA must be cut with the same restriction enzyme used to cut the human DNA.
    c. The fragment must be joined to other fragments containing the same gene so many copies can be made.
    d. Plasmid DNA must be joined with viral DNA to make a vector.
    e. The fragment must be introduced into a bacterial cell.

24. What is the next step in a genetic engineering project after a recombinant DNA molecule has been made?
    a. The recombinant DNA molecule must be cut with restriction enzymes.
    b. The recombinant DNA molecule must be cut with ligase.
    c. The recombinant DNA molecule must be cloned.
    d. The recombinant DNA molecule must be separated by electrophoresis.
    e. The recombinant DNA molecule must be used to produce "sticky ends."

25. Which describes a recombinant DNA molecule in a bacterial cell?
    a. It must contain a human gene.
    b. It must be made of plasmid DNA.
    c. It must contain a prokaryotic promoter.
    d. It must contain prokaryotic RNA polymerase.
    e. It must be cut with a restriction enzyme.

26. Which describes a recombinant DNA molecule in a bacterial cell?
    a. It must be introduced into the bacterial cell by a virus.
    b. It must not have any introns.
    c. It must have a eukaryotic operator region.
    d. It must have a eukaryotic DNA polymerase.
    e. It must have been cloned prior to entry into the bacterial cell.

27. What is a function of gel electrophoresis?
    a. It is used to produce restriction fragments.
    b. It is used to separate restriction fragments.
    c. It is used to clone recombinant DNA molecules.
    d. It is used to repair fragmented recombinant DNA molecules.
    e. It is used to determine the function of the cloned gene.

28. What is the function of the radioactive probe used after gel electrophoresis?
    a. It is used to locate a DNA fragment containing a specific gene.
    b. It is used to join DNA fragments together.
    c. It is used to produce restriction fragments.
    d. It is used to clone recombinant DNA molecules.
    e. It is used to determine the nucleotide sequence of a restriction fragment.

29. Which describes the polymerase chain reaction (PCR)?
    a. RNA polymerase from hot springs bacteria is used.
    b. Protein primers must be used.
    c. It is a form of cloning.
    d. The reaction must occur within a bacterial cell.
    e. Strand separation is accomplished by an enzyme.

30. Which describes the polymerase chain reaction (PCR)?
    a. PCR requires large starting amounts of DNA.
    b. Only synthetic DNA molecules can be used in PCR.
    c. Only natural DNA molecules can be used in PCR.
    d. The DNA is alternately heated, then cooled during PCR.
    e. PCR occurs rapidly, even in the absence of enzymes.

## Concept Map Construction

Construct a concept map for each group of terms. Be sure to include appropriate connector phrases. You may add other terms as necessary and use terms in the singular or plural form.

1. polymerase chain reaction, DNA cloning, vector, DNA polymerase, anabolic reaction
2. restriction enzyme, plasmid, intron, sticky end, protein
3. genetic marker, chorionic villus sampling, autosomal recessive allele, amniocentesis, DNA

# Chapter 12

# Plant Life: Form and Function

## Go Figure!

Figure 12-2: In what way are the root and shoot systems related to the dermal tissue, ground tissue, and vascular tissue systems?

Figure 12-3: In what way are apical meristems similar to secondary (lateral) meristems?

How do they differ?

Figure 12-4: The stem in this illustration was cut in cross section. In what way must the stem be cut in order to see what the cells look like lengthwise?

Why are the vascular tissues not discussed in this figure?

Figure 12-5: To what tissue do sclereids belong?

Figure 12-6: What type of tissue comprises the bulk of the ground tissue system?

Figure 12-7: How do guard cells enable a plant to maintain homeostasis during water stress?

Figure 12-8: In what ways are lenticels and stomates similar?

How are they different?

Figure 12-11: Why doesn't secondary phloem accumulate to the same degree as secondary xylem?

Which tissues are included in bark?

Which tissue is "bark" to someone not trained in botany?

Figure 12-12: Since 90% of the water lost by a plant exits through the stomates, why don't plants keep them closed all the time and minimize their loss?

Figure 12-13: What are the conducting cells of xylem called?

Are they alive or dead when functional?

What structure do they form when lined up end to end?

How do growth and metabolism contribute to the pressure gradient necessary for xylem transport?

Figure 12-15: Why are aphids considered pests by many plant growers?

Figure 12-16: Are the tissues at the top of Figure 12-16a acting as sources or sinks?

How can storage tissue act both as a source and a sink?

Figure 12-18: Which plant group is characterized by a tap root system?

Which plant group is characterized by a fibrous (adventitious) root system?

Figure 12-21: Which tissue system gives rise to the pericycle?

Why do you suppose lateral roots arise from within the root rather than from the surface?

Figure 12-22: What physical characteristic makes fungi efficient water absorbers?

Suggest reasons why 100% of the species in the North Carolina forest study didn't show evidence of mycorrhizal linkages.

Figure 12-23: Would water enter roots as efficiently if minerals were extracted from the air rather than the soil? Defend your answer.

Observe the endodermal cells. Each cell has six walls: one on each side, an upper and lower, and a front and back. The figure shows the Casparian strip in the two side walls. Which other walls, if any, must also contain a Casparian strip to produce an effective barrier?

What purpose does the Casparian strip serve?

What are the connections between cortical cells called?

Figure 12-24: Why does guttation only occur at night?

Figure 12-25: Nitrogen gas ($N_2$) comprises almost 80% of air. Why are nitrogen-fixing bacteria essential to the survival of plants in nitrogen-poor soils?

Figure 12-26: Notice that stem and root xylem is towards the center and phloem is towards the surface (see Figures 12-17 and 12-21). Propose an anatomical reason why leaf xylem is towards the upper leaf surface and phloem is towards the lower leaf surface.

## Matching Questions

Write the letter of the phrase that best matches the numbered term on the left. Use each only once.

____ 1. monocot

____ 2. dicot

____ 3. cuticle

____ 4. vascular cambium

____ 5. apical meristem

____ 6. vessel member

____ 7. sieve-tube member

____ 8. Casparian strip

____ 9. parenchyma

____10. collenchyma

____11. vascular tissue system

____12. dermal tissue system

____13. xylem

____14. phloem

____15. sink

____16. source

____17. stomates

____18. transpiration

____19. root nodule

____20. mycorrhizae

____21. phloem loading

____22. phloem unloading

____23. plasmodesmata

a. characterized by parallel veins in leaves

b. conducting cell of xylem

c. constitutes the cellular covering of a plant

d. conducting cell of phloem

e. the most abundant tissue in herbaceous plants

f. characterized by flower parts in fours or fives

g. waxy covering on the epidermis

h. responsible for secondary growth in a plant

i. includes all the conducting cells of a plant

j. found in the walls of endodermal cells

k. support tissue of the primary plant body

l. region of a plant responsible for primary growth

m. process of evaporative water loss

n. tissue responsible for organic molecule transport

o. occurs at the sink

p. tissue responsible for mineral and water transport

q. connections between the cytoplasm of plant cells

r. fungi associated with roots of most vascular plants

s. growing plant parts or storage tissues

t. primary site of water loss by a plant

u. occupied by nitrogen-fixing bacteria

v. occurs at the source

w. photosynthetic or storage tissues

## Multiple Choice Questions

1. Which term is applied to the underground portion of a plant?
   a. ground system
   b. dermal system
   c. vascular system
   d. root system
   e. shoot system

2. Which term is applied to the aerial portion of a plant?
   a. ground system
   b. dermal system
   c. vascular system
   d. root system
   e. shoot system

3. Which DOES NOT describe monocots?
   a. parallel leaf venation
   b. flower parts in 4s and 5s
   c. scattered vascular bundles in stem
   d. fibrous root system
   e. no secondary growth

4. Which DOES NOT describe dicots?
   a. two types of photosynthetic cells in leaves
   b. tap root system
   c. pith in center of root
   d. two cotyledons in embryo
   e. stem vascular bundles in rings

5. Which DOES NOT describe primary growth?
   a. It produces primary tissues.
   b. It causes elongation of the plant.
   c. It involves an apical meristem.
   d. It occurs only in annual plants.
   e. It produces vascular and other tissues.

6. Which DOES NOT describe secondary growth?
   a. It produces secondary tissues.
   b. It involves vascular cambium.
   c. It involves cork cambium.
   d. It increases plant diameter.
   e. It produces only simple tissues.

7. Which DOES NOT describe parenchyma?
   a. Parenchyma cells are versatile in function.
   b. Parenchyma cells have a thick secondary cell wall.
   c. It may combine with other tissues to form a complex tissue.
   d. It is a common tissue in herbaceous plants.
   e. It is often involved in photosynthesis.

8. Which DOES NOT describe collenchyma?
   a. The cells are spherical.
   b. The cells often have unevenly thickened cell walls.
   c. It physically supports growing parts of the plant.
   d. The cells perform photosynthesis.
   e. It lies just below the epidermis in stems and leaves.

9. Which describes sclerenchyma fibers?
   a. They constitute a complex tissue.
   b. They have thin primary and secondary walls.
   c. They are long and thin.
   d. The secondary wall is soft to allow growth.
   e. The cells are only functional when living.

10. Which describes sclereids?
    a. They are a type of collenchyma cell.
    b. They have hard, thick secondary walls.
    c. They typically are long and thin.
    d. They come in only one shape.
    e. The cells are only functional when living.

11. Which DOES NOT describe epidermis?
    a. It is the cellular covering of the primary plant body.
    b. The epidermal cells of the shoot system are covered with a waxy cuticle.
    c. It contains openings to the interior of the plant called stomates.
    d. It belongs to the dermal tissue system.
    e. Its primary function is transport of material between parts of the plant.

12. Which DOES NOT describe peridermis?
    a. It contains openings to the interior of the plant called stomates.
    b. It is the cellular covering of the secondary plant body.
    c. It is a component of bark.
    d. It belongs to the dermal tissue system.
    e. It contains suberized cork cells.

13. Which DOES NOT describe xylem?
    a. It belongs to the vascular tissue system.
    b. The conducting cells are functional when dead.
    c. It transports water and minerals.
    d. Secondary xylem is also called bark.
    e. Vessel members and tracheids are the conducting cells of xylem.

14. Which DOES NOT describe phloem?
    a. The conducting cells are functional when dead.
    b. The conducting cells are called sieve-tube members.
    c. It transports photosynthetic products.
    d. It belongs to the vascular tissue system.
    e. The conducting cells are located next to companion cells.

15. Which belongs to the ground tissue system?
    a. vascular cylinder
    b. epidermis
    c. cortex
    d. cork
    e. pericycle

16. Which belongs to the ground tissue system?
    a. pith
    b. xylem
    c. phloem
    d. companion cells
    e. guard cells

17. Which DOES NOT describe a primary stem?
    a. There is a meristem at the tip.
    b. There is an active cambium.
    c. Some growth is due to lengthening of cells.
    d. It may belong to a dicot.
    e. It may belong to a monocot.

18. Which DOES NOT describe a plant in secondary growth?
    a. There is an active cambium.
    b. It may be a monocot.
    c. It may be a dicot.
    d. There is an active meristem at the tips of each shoot.
    e. Its secondary xylem is also known as wood.

19. Roots arising from stem tissue are called
    a. storage roots.
    b. tap roots.
    c. root hairs.
    d. adventitious roots.
    e. lateral roots.

20. Where are root hairs located?
    a. on the root cap
    b. at the apical meristem
    c. at the region of elongation
    d. at the region of maturation
    e. at the Casparian strip

21. What is the function of root hairs?
    a. They increase surface area for water and mineral absorption.
    b. They increase surface area for anchorage of the plant.
    c. They increase volume of the root for storage.
    d. They add cells to the root meristem.
    e. They form a waterproof barrier on the surface of the root.

22. What is the function of endodermis?
    a. It insures that water and minerals only enter the vascular cylinder through living cells.
    b. It increases surface area for absorption.
    c. It gives rise to lateral roots.
    d. It acts as a storage tissue.
    e. It performs photosynthesis.

23. What is the flat part of a leaf called?
    a. node
    b. internode
    c. petiole
    d. blade
    e. axillary bud

24. Which tissue is found in the central portion of a leaf?
    a. epidermis
    b. mesophyll
    c. endodermis
    d. guard cells
    e. peridermis

25. Which tissue is responsible for water and mineral transport up a vascular plant?
    a. parenchyma
    b. epidermis
    c. peridermis
    d. xylem
    e. phloem

26. Which tissue is responsible for organic compound transport throughout a vascular plant?
    a. parenchyma
    b. epidermis
    c. peridermis
    d. xylem
    e. phloem

27. Which describes the process of water absorption by vascular plants?
    a. Water is actively transported into roots.
    b. Water moves only through living cells in getting to the vascular tissue.
    c. Water movement is along its concentration gradient.
    d. Water movement is independent of mineral movement.
    e. Water movement is restricted in roots by the cuticle.

28. Which describes the process of mineral absorption by vascular plants?
    a. Minerals are actively transported into roots.
    b. Minerals are destined for phloem tissue in the root.
    c. Minerals typically move along their concentration gradient.
    d. Minerals are concentrated in epidermal cells to produce a concentration gradient favorable to mineral movement into the xylem.
    e. Plasmodesmata are too small to allow passage of macronutrients.

29. Which is a function of potassium (K)?
    a. It is a major component of organic molecules.
    b. It helps regulate stomatal opening and closing.
    c. It acts as an enzyme cofactor.
    d. It acts as a coenzyme.
    e. It is an electron acceptor in photosynthesis.

30. Which is a function of iron (Fe)?
    a. It is a component of cytochromes.
    b. It is a component of the middle lamella.
    c. It helps regulate stomatal opening and closing.
    d. It is a major component of ATP.
    e. It a component of Coenzyme A.

31. Which describes xylem transport?
    a. Most materials are pushed up the stem by root pressure.
    b. Materials are moved primarily by guttation.
    c. Xylem transport occurs at a uniform rate throughout the day.
    d. Cohesion of water molecules is necessary for water transport.
    e. Xylem transport requires cellular active transport.

32. Which describes xylem transport?
    a. Evaporation of water from leaves results in movement up the xylem.
    b. Pressure from carbohydrate absorption pushes materials up the xylem.
    c. It requires facilitated diffusion of minerals and water follows by osmosis.
    d. It involves pressure produced from breaks in the water column in xylem vessels.
    e. Loading of xylem vessels is by endocytosis.

33. Which should result in faster transpiration?
    a. increased humidity
    b. increased temperature
    c. decreased soil surface
    d. decreased root surface
    e. decreased pressure within guard cells

34. Which should result in faster transpiration?
    a. increased breeze
    b. increased soil surface
    c. increased ATP hydrolysis
    d. decreased leaf surface
    e. decreased light

35. Which DOES NOT describe phloem transport?
    a. Phloem loading occurs at the source.
    b. Movement is due to positive pressure in sieve tubes.
    c. Pressure is higher at the sink than at the source.
    d. ATP hydrolysis is required at the source.
    e. Flow is from source to sink.

36. Which DOES NOT describe phloem transport?
    a. ATP hydrolysis is required at the sink.
    b. Sucrose is the main organic compound transported.
    c. A pressure gradient exists in sieve tubes.
    d. The rate is regulated by guard cells.
    e. Phloem unloading occurs at the sink.

37. How are phloem and xylem transport different?
    a. Xylem transport requires cellular energy; phloem transport does not.
    b. Xylem transport is exclusively up the stem; phloem transport is exclusively down the stem.
    c. Xylem transport occurs in cells that are dead; phloem transport occurs in cells that are technically alive.
    d. Xylem transport occurs as a result of positive pressure; phloem transport occurs as a result of negative pressure.
    e. Xylem transport occurs only in roots; phloem transport occurs only in leaves.

38. How are phloem and xylem transport different?
    a. Phloem transport requires cellular energy; xylem transport does not.
    b. Phloem transport is exclusively up the stem; xylem transport is exclusively down the stem.
    c. Phloem transport occurs in cells that are dead; xylem transport occurs in cells that are alive.
    d. Phloem transport occurs as a result of suction; xylem transport occurs as a result of pushing.
    e. Phloem transport occurs only in stems; xylem transport occurs only in roots.

## Concept Map Construction

Construct a concept map for each group of terms. Be sure to include appropriate connector phrases. You may add other terms as necessary and use terms in the singular or plural form.

1. lenticel, epidermis, ground tissue system, mesophyll, Photosystem I
2. transpirational pull, stomate, vessel member, evaporation, root pressure
3. ATP, source, companion cell, cohesion, phloem loading

# Chapter 13

# Plant Life: Reproduction and Development

## Go Figure!

Figure 13-2: All parts of the flower are thought to be modified leaves. Suggest ways in which hollow structures such as the pistil and anthers could be formed from flat leaves.

Figure 13-4: Why do feathery stigmas and production of large amounts of pollen improve chances of successful pollination in wind pollinated flowers?

Figure 13-5: What process produces megaspores in the plant life cycle?

What process produces egg gametes in the plant life cycle?

What structure represents the female gametophyte?

What is the ploidy of the female gametophyte?

How do spores differ from gametes? How are they similar?

Figure 13-6: What process produces microspores in the plant life cycle?

What process produces sperm in the plant life cycle?

What structure represents the male gametophyte?

What is the ploidy of the male gametophyte?

Figure 13-7: Distinguish between the processes of pollination and fertilization.

Figure 13-8: What process makes endosperm tissue multicellular?

How does the single celled zygote give rise to the distinctly different embryo and suspensor?

Besides reading the label, how could you tell this is a diagram of a dicot embryo?

Figure 13-9: Besides reading the label, how could you tell this is a diagram of a monocot embryo?

How many ovules are in each ovary of the corn plant?

Figure 13-10: Most people eat apples with the "stem" on the top. Relative to the flower that produced the fruit, do we eat apples right side up or upside down?

Figure 13-11: Compare the third seedling drawing (second from right) to the final early embryo drawing shown in Figure 13-8. Notice all parts of the seedling are present in the embryo depicted at the lower left (Figure 13-8f). Seedling production simply involves differential growth of the embryonic root and shoot at the expense of the cotyledons. What important function have the cotyledons performed between Figure 13-8f and the first drawing in Figure 13-11?

Why is it incorrect to say the epicotyl is solely responsible for producing the stem of the seedling?

Why do the cotyledons wither soon after germination?

Figure 13-12: Compare the corn seedling drawing to the embryo shown in Figure 13-9. Notice all parts of the seedling are present in the embryo. Germination simply involves growth of the embryo (the corn kernel stays the same size).

What is the role of the corn cotyledon?

How is its role different from dicot cotyledons?

What other difference between monocot and dicot cotyledons is evident in Figures 13-11 and 13-12?

Figure 13-14: What does IAA stand for?

To which class of hormone does IAA belong?

Why do gardeners pinch the tips of stems to produce fuller growth in their plants?

Figure 13-15: What three factors may be involved in timing growth with respect to season?

Figure 13-16: Why did botanists conclude that dark period was more important than light period in promoting flowering? (Hint: compare the bar graphs for poinsettia and hibiscus.)

Figure 13-17: Which tropism is illustrated by the pea and the Virginia creeper?

## Matching Questions

Write the letter of the phrase that best matches the numbered term on the left. Use each only once.

____ 1. pollination
____ 2. anther
____ 3. stigma
____ 4. endosperm mother cell
____ 5. generative cell
____ 6. mitosis
____ 7. meiosis
____ 8. megaspore
____ 9. imperfect flower
____10. complete flower
____11. female gametophyte
____12. male gametophyte
____13. ovary
____14. ovule
____15. zygote
____16. germination
____17. gibberellin
____18. ethylene gas
____19. abscisic acid
____20. long-day plant
____21. short-day plant
____22. gravitropism
____23. thigmotropism

a. pollen producing part of the flower
b. a flower that lacks stamens or pistils
c. develops into the embryo
d. flowering occurs when days are longer than some critical minimum length
e. site where pollen lands
f. process that produces spores
g. a perfect flower with sepals and petals
h. mobilizes food resources in the seed
i. process of pollen transfer between flowers
j. develops into the embryo sac
k. develops into a fruit
l. flowering occurs when day length is shorter than some critical maximum length
m. cell that produces sperm
n. a response to gravity
o. first stage of seedling development
p. a growth response initiated by contact with another object
q. develops into a seed
r. a cell with two haploid nuclei
s. promotes fruit ripening
t. process that produces gametes in plants
u. inhibits cell division leading to plant dormancy
v. the embryo sac
w. the pollen grain

## Multiple Choice Questions

1. Which is the process of pollen transfer?
   a. germination
   b. fertilization
   c. pollination
   d. parthenocarpy
   e. dormancy

2. Which is the process of gamete fusion?
   a. germination
   b. fertilization
   c. pollination
   d. parthenocarpy
   e. dormancy

3. Which DOES NOT describe flower structure?
   a. Large, showy petals attract pollinators.
   b. The stigma produces the female gametophyte.
   c. All parts of the flower are thought to be modified leaves.
   d. The anthers produce male gametophytes.
   e. The flower parts are attached to the receptacle.

4. Which DOES NOT describe flower structure?
   a. The pistil is made of modified leaves called carpels.
   b. Sepals protect the developing flower.
   c. Stamens are the male part of the flower.
   d. The pistil is the female part of the flower.
   e. Many eggs develop within the embryo sac.

5. *Gossypium hirsutum* is the cotton plant. Its flowers occur singly and are positioned above three leaf-like bracts. The calyx is small and cup-shaped, and composed of five sepals. The rest of the flower consists of a corolla with five petals, numerous stamens forming a tube around the style, and an ovary with three to five carpels, each with several ovules. Which describes *Gossypium hirsutum*?
   a. The flowers are imperfect.
   b. The flowers are incomplete.
   c. *Gossypium hirsutum* is a dicot.
   d. *Gossypium hirsutum* uses flowers to reproduce asexually.
   e. Each fruit should contain one seed.

6. *Papaver somniferum* is the poppy plant. It is the source of opium and morphine, as well as the poppy seeds used by the baking industry. Its flowers are borne singly and consist of two sepals that are deciduous (they drop off), four petals, numerous stamens, and an ovary made of many fused carpels containing many ovules. Which describes *Papaver somniferum*?
   a. The flowers are imperfect.
   b. The flowers are incomplete.
   c. *Papaver somniferum* is a monocot.
   d. *Papaver somniferum* uses flowers to reproduce asexually.
   e. Each fruit should contain many seeds.

7. Which is the immediate product of meiosis in the pistil of a flower?
   a. sperm
   b. egg
   c. zygote
   d. spore
   e. embryo sac

8. Which is the immediate product of meiosis in the anther of a flower?
   a. sperm
   b. egg
   c. spore
   d. zygote
   e. embryo sac

9. What is another name for the male gametophyte?
   a. embryo sac
   b. anther
   c. endosperm
   d. pollen
   e. spore

10. What is another name for the female gametophyte?
   a. embryo sac
   b. anther
   c. endosperm
   d. pollen
   e. spore

11. Which describes the embryo sac?
   a. It consists of eight cells and eight nuclei.
   b. All the cells are diploid.
   c. The generative cell produces the egg.
   d. It contains the egg.
   e. It develops from the egg.

12. Which describes the embryo sac?
   a. It consists of two cells.
   b. All the cells are haploid.
   c. The tube cell receives the pollen tube during fertilization.
   d. It is contained within the anther.
   e. It develops from a spore.

13. Which describes the pollen grain?
   a. It consists of seven cells and eight nuclei.
   b. All the cells are haploid.
   c. It has an opening called the micropyle.
   d. The endosperm cell is triploid.
   e. It develops from a sperm cell.

14. Which describes the pollen grain?
   a. The tube cell is diploid.
   b. The generative cell is diploid.
   c. It develops into a spore.
   d. The generative cell produces two sperm cells.
   e. Sperm escape the pollen grain through its micropyle.

15. Which DOES NOT describe a dicot seed?
   a. The embryo has two cotyledons.
   b. Each cotyledon contains endosperm.
   c. The radicle is part of the epicotyl.
   d. It developed from an ovule.
   e. The seed coats developed from the outer ovule layers.

16. Which DOES NOT describe a monocot seed?
    a. Each cotyledon contains endosperm.
    b. The embryo has one cotyledon.
    c. It developed from an ovule.
    d. The epicotyl develops into most of the shoot system.
    e. Endosperm is triploid tissue.

17. Each fruit of an orange tree comes from the ripened ovary of a single flower. How is the orange fruit classified?
    a. simple fruit
    b. aggregate fruit
    c. multiple fruit
    d. vegetative fruit
    e. polycarpic fruit

18. The blackberry is composed of single-seeded, fleshy parts derived from a single flower. How is the blackberry fruit classified?
    a. simple fruit
    b. aggregate fruit
    c. multiple fruit
    d. vegetative fruit
    e. parthenocarpic fruit

19. Which describes bean seedling development?
    a. Its development is representative of monocots.
    b. The radicle is the first embryonic part to emerge from the seed.
    c. The radicle provides a source of food until the seedling begins photosynthesis.
    d. The hypocotyl is straight until the shoot breaks through the soil surface.
    e. The cotyledons form after the seedling becomes photosynthetic.

20. Which describe corn seedling development?
    a. The cotyledon absorbs nutrients from the seed coat.
    b. The epicotyl develops into the root system.
    c. Corn seedling development is representative of dicots.
    d. After emergence of the epicotyl, the cotyledons wither and fall off.
    e. The seedling becomes autotrophic after emergence of the epicotyl and development of foliage leaves.

21. According to your text, which is NOT an action of auxin?
    a. It stimulates axillary bud development.
    b. It stimulates elongation of cells.
    c. It is involved in the phototropic response.
    d. It promotes flowering.
    e. It stimulates mitosis in vascular cambium.

22. According to your text, which is NOT an action of gibberellin?
    a. It promotes digestion of endosperm.
    b. It promotes pollen tube development.
    c. It makes fruit grow larger.
    d. It promotes dormancy.
    e. It stimulates flower formation.

23. According to your text, which is an action of ethylene gas?
    a. It is involved in the phototropic response.
    b. It promotes fruit ripening.
    c. It stimulates production of chlorophyll.
    d. It stimulates rapid cell division in roots.
    e. It inhibits flower production.

24. According to your text, which is an action of abscisic acid?
    a. It promotes elongation of stem cells not receiving direct illumination.
    b. It promotes elongation of stem cells receiving direct illumination.
    c. It stimulates fruit ripening.
    d. It prolongs the life of leaves and flowers.
    e. It promotes dormancy.

25. Which is used by plants to coordinate their development with environmental changes?
    a. temperature
    b. humidity
    c. insect pollinators
    d. wind
    e. spines

26. Which is used by plants to coordinate their development with environmental changes?
    a. fruit production
    b. waxy cuticle
    c. photoperiod
    d. photo opportunity
    e. root hair production

*NOTE: Answers for questions 27 and 28 are provided in the Appendix. Questions 29 and 30 constitute a similar sequence, but the answers are not given.*

27. The poinsettia is a short-day plant. Which describes poinsettias?
    a. They only flower if days are short and nights are long.
    b. They only flower if days are shorter than nights.
    c. They only flower if days are shorter than some critical maximum length.
    d. They only flower if days are longer than some critical minimum length.
    e. They flower regardless of day length.

28. The poinsettia is a short-day plant. Which describes poinsettias?
    a. They flower in February.
    b. They flower in spring.
    c. They flower in summer.
    d. They flower in autumn.
    e. They flower in response to cold temperatures.

29. Clover is a long-day plant. Which describes clover?
    a. It only flowers if days are long and nights are short.
    b. It only flowers if days are longer than nights.
    c. It only flowers if days are longer than some critical minimum length.
    d. It only flowers if days are shorter than some critical maximum length.
    e. It flowers regardless of day length.

30. Clover is a long-day plant. Which describes clover?
    a. It flowers in September.
    b. It flowers in November.
    c. It flowers in spring.
    d. It flowers in winter.
    e. It flowers in response to cold temperatures.

## Concept Map Construction

Construct a concept map for each group of terms. Be sure to include appropriate connector phrases. You may add other terms as necessary and use terms in the singular or plural form.

1. petal, incomplete flower, perfect flower, ovule, leaf
2. apical dominance, DNA, hormone, gibberellin, ovary
3. seed, cotyledon, florigen, chloroplast, long-day plant

# Chapter 14

# Animal Form and Function

## Go Figure!

Figure 14-1: Why do you suppose simple squamous epithelium is in the lungs?

Why do you suppose stratified squamous epithelium comprises the outer skin (epidermis)?

Figure 14-2: Cite characteristics that suggest blood is a connective tissue.

Figure 14-3: Why do you think the authors of your text have chosen to introduce muscle *and* nerve tissue in the same illustration?

Figure 14-7: What is meant by the statement that the stapes "can be *traced* to a bone in the skull of ancestral fishes"?

## Matching Questions

Write the letter of the phrase that best matches the numbered term on the left. Use each only once.

____ 1. homeostasis

____ 2. receptor

____ 3. negative feedback mechanism

____ 4. human body effectors

____ 5. positive feedback mechanism

____ 6. tissue

____ 7. epithelial tissue

____ 8. epidermis

____ 9. connective tissue

____10. extracellular matrix

____11. collagen

____12. cornea

____13. blood vessel constriction

____14. muscle tissue

____15. nerve tissue

____16. organ

____17. immune system

____18. hormones

a. a change in one variable accelerates further change in that variable

b. epithelium forming outer layer of human skin

c. a sheet of cells lining body spaces

d. tissue specialized for contraction

e. a change in one variable reduces further change in that variable

f. connective tissue protein

g. a function of muscle tissue

h. an organ that monitors external or internal conditions

i. general name for any group of cells with a related structure and common function

j. maintenance of conditions within narrow limits

k. connective tissue covering the eye

l. includes muscles and glands

m. loosely organized cells within a nonliving matrix

n. chemicals produced by the endocrine system of animals

o. includes very long cells along which an electrical signal is sent

p. discrete structure, such as the brain, composed of tissues

q. protects an animal from foreign substances which have entered its body

r. general name of nonliving material comprising the bulk of most connective tissues

## Multiple Choice Questions

1. The human body maintains temperature homeostasis by
   a. a positive feedback mechanism.
   b. a negative feedback mechanism.
   c. the osmoregulatory system.
   d. the immune system.
   e. temperature receptors which lower body temperature when it is too high.

2. The human body regulates blood clotting by means of
   a. a positive feedback mechanism.
   b. a negative feedback mechanism.
   c. the excretory system.
   d. an effector which lowers clotting rate after clotting has begun.
   e. a receptor which lowers clotting rate when it is too high.

3. The human body receives information about homeostatically maintained internal and external conditions through
   a. positive feedback mechanisms.
   b. negative feedback mechanisms.
   c. receptors.
   d. body integrators.
   e. effectors.

4. To maintain homeostasis, the human body processes incoming information and sends signals to the body from
   a. positive feedback mechanisms.
   b. negative feedback mechanisms.
   c. sensors.
   d. body integrators.
   e. effectors.

5. The lowest/simplest organizational level composed of a group of cells with related structure and function best describes
   a. a tissue.
   b. an organ.
   c. an organ system.
   d. an antibody.
   e. extracellular material.

6. An organized aggregate of cell groups with different structures and functions working together to perform a particular task best describes
   a. a tissue.
   b. an organ.
   c. an organ system.
   d. an antibody.
   e. biochemicals.

7. A sheet of tightly adhering cells best describes
   a. epithelial tissue.
   b. connective tissue.
   c. muscle tissue.
   d. nervous tissue.
   e. collagen.

8. A group of loosely organized cells surrounded by a nonliving extracellular matrix best describes
   a. epithelial tissue.
   b. connective tissue.
   c. muscle tissue.
   d. nervous tissue.
   e. collagen.

9. A group of cells containing elongated protein filaments which allow each cell to contract best describes
   a. epithelial tissue.
   b. connective tissue.
   c. muscle tissue.
   d. nervous tissue.
   e. collagen.

10. A group of cells capable of sending a "signal" along their membrane to other cells best describes
    a. epithelial tissue.
    b. connective tissue.
    c. muscle tissue.
    d. nervous tissue.
    e. collagen.

11. Which is a function of epithelial tissue?
    a. cell contraction to effect movement of the organism
    b. formation of an intracellular matrix composed of protein and polysaccharide
    c. protection of underlying tissues
    d. communication between different areas of an animal's body
    e. blood formation

12. Which is a function of epithelial tissue?
    a. changing the shape of the human eye lens
    b. secretion of materials into body spaces
    c. movement of food through the digestive tract
    d. formation of the lower layer of the skin (dermis)
    e. constriction of blood vessels

13. Collagen fibers are found in most
    a. epithelial tissues.
    b. connective tissues.
    c. muscle tissues.
    d. nervous tissues.
    e. tissues.

14. Which IS NOT made primarily of connective tissue?
    a. bone
    b. adipose
    c. cornea
    d. tendon
    e. endocrine gland

15. Functions of muscle tissue include
    a. formation of the outer layer of the eyeball.
    b. formation of an intracellular matrix composed of protein and polysaccharide.
    c. movement of food through the digestive tract.
    d. communication between different areas of an animal's body.
    e. formation of blood.

16. Functions of muscle tissue include
    a. production of hormones.
    b. production of digestive enzymes.
    c. constriction of blood vessels.
    d. formation of collagen.
    e. transport of fat to adipose tissue.

17. Functions of a urinary system include
    a. absorption of macromolecules produced by digestion.
    b. ridding the body of metabolic waste products.
    c. maintenance of body shape.
    d. antibody production.
    e. oxygen absorption.

18. Functions of a urinary system include
    a. expulsion of carbon dioxide gas.
    b. transport of oxygen gas.
    c. storage of gametes until they are needed.
    d. regulation of salt and water content.
    e. providing rigid structures upon which muscles can exert a force.

19. Functions of a skeletal system include
    a. absorption of macromolecules produced by digestion.
    b. ridding the body of metabolic waste products.
    c. maintenance of body shape.
    d. antibody production.
    e. oxygen absorption.

20. Functions of a skeletal system include
    a. expulsion of carbon dioxide gas.
    b. transport of oxygen gas.
    c. regulation of sodium chloride (table salt) and water content.
    d. providing rigid structures upon which muscles can exert a force.
    e. production of the body's hormones.

21. Which system is primarily involved in transport of materials throughout the body?
    a. excretory system
    b. digestive system
    c. circulatory system
    d. reproductive system
    e. respiratory system

22. Which system is primarily involved in gas exchange with the environment?
    a. excretory system
    b. digestive system
    c. circulatory system
    d. reproductive system
    e. respiratory system

23. Which system is primarily involved in protecting the body from foreign substances?
    a. muscular system
    b. digestive system
    c. immune system
    d. integumentary system
    e. skeletal system

24. An animal shifting the position of its entire body from one location to another is called
    a. movement.
    b. an appendage.
    c. locomotion.
    d. an antibody.
    e. a receptor.

## Concept Map Construction

Construct a concept map for each group of terms. Be sure to include appropriate connector phrases. You may add other terms as necessary and use terms in the plural or singular form.

1. homeostasis, negative feedback mechanism, hormone, Golgi body, collagen
2. ribosome, contractile protein, tissue, epidermis, organ
3. plasma membrane, collagen, digestive system, respiratory system, receptor

# Chapter 15

# Regulating Bodily Functions: The Nervous and Endocrine Systems

## Go Figure!

Figure 15-1: Assume a neural impulse begins at a dendrite and travels to a target cell. Name each neuron component over which it travels in reaching its goal.

What is found at the end of each axon collateral fiber?

Using just the information provided in this diagram and the preceding chapter, what do you think the target cell (muscle) response will be?

Is the myelin sheath part of the nerve cell?

Figure 15-2: Identify the motor neuron cell body. Is the axon to the left or right of the cell body?

Identify the interneuron cell body on the interneuron to the right. Is the axon above or below the cell body?

Figure 15-3: In what ways does repolarization differ from polarization?

Draw an arrow to indicate the direction of impulse propagation.

Figure 15-4: This illustration shows skeletal muscle cells as the target of this neuron. What other types of cells could be the target of a neuron?

Figure 15-7: Is the cell body of the sensory neuron found within the spinal cord?

Which part of the sensory neuron acts as the receptor? Which part enters the spinal cord?

Is the cell body of the motor neuron found within the spinal cord?

Which part of the motor neuron carries an impulse to the muscle?

Figure 15-11: Why isn't there a ventral root ganglion?

Cerebrospinal fluid is found between what two anatomical structures?

What bony structures surround and protect the spinal cord? Given this association, what might cause a "pinched" spinal nerve?

Figure 15-12: During times of stress, what advantage does an organism gain by sympathetic stimulation reducing salivary gland secretions and inhibiting stomach and intestine muscular contractions?

What are the advantages of increasing the release of glucose from the liver during times of stress?

Figure 15-15: In what way does the anterior pituitary capillary arrangement differ from capillary structure in the posterior pituitary? Base your answer only upon the figure.

What type of cell is present in the anterior pituitary and missing from the posterior pituitary?

From the illustration, what is the first event in the series leading to release of hormone from both parts of the pituitary?

Figure 15-17: What type of macromolecule do you think the receptor might be?

How might cAMP affect an enzyme to convert it from an inactive state to an active one?

What are the molecules represented by a yellow circle and two gray lines?

What does the blue circle represent in Figure 15-17b? What do the light blue parts represent?

## Matching Questions

Write the letter of the phrase that best matches the numbered term on the left. Use each only once.

____ 1. neuron

____ 2. cell body

____ 3. dendrite

____ 4. nervous impulse

____ 5. target cell

____ 6. sensory neuron

____ 7. node of Ranvier

____ 8. leak channel

____ 9. depolarization

____10. acetylcholine

____11. cortisol

____12. insulin

____13. neurosecretory cells

____14. posterior pituitary gland

____15. releasing factor

____16. antidiuretic hormone

____17. acromegaly

____18. mineralocorticoids

____19. adrenal gland

____20. adrenalin

____21. thyroid gland

____22. hypothyroidism

____23. Type II diabetes

a. narrow cytoplasmic process which conducts neural impulses toward a nerve cell's cell body

b. abnormally low levels of thyroid hormone

c. a decrease in polarization

d. in the neck near the trachea

e. a gland, neuron, or muscle cell that responds to a neural impulse

f. also known as epinephrine

g. general name for a nerve cell

h. aldosterone is one of these

i. contains the nucleus of a nerve cell

j. results from too much growth hormone

k. stimulative product of hypothalamic neurosecretory cells not stored in the posterior pituitary

l. glucocorticoid hormone of the adrenal cortex

m. produced by cells originating in the hypothalamus and reduces water loss from the kidneys

n. wave of depolarization traveling along a nerve cell

o. results from insulin receptor abnormalities

p. intramembrane passage that is always open

q. stores hormones produced by cells originating in the hypothalamus

r. nerve cell that carries impulses towards the CNS

s. look like neurons, function like endocrine cells

t. a neurotransmitter

u. one found on top of each kidney

v. uninsulated gap of a myelinated axon

w. hormone produced in the pancreas

## Multiple Choice Questions

1. The human nervous system
   a. regulates heart rate.
   b. removes metabolic wastes.
   c. consists of modified epithelial cells.
   d. contains a single type of cell.
   e. forms the lining of the body cavities.

2. The human nervous system
   a. forms the outer covering of the eye.
   b. helps coordinate homeostasis.
   c. relies only on hormones for sending messages.
   d. is composed of molecular subunits called nodes of Ranvier.
   e. is limited to the brain and spinal cord.

3. Target cells of the human body include
   a. dendrites.
   b. myelin.
   c. nodes of Ranvier.
   d. muscle cells.
   e. connective tissue cells.

4. Target cells of the human body include
   a. cerebrospinal fluid.
   b. neurons.
   c. synaptic knobs.
   d. axons.
   e. epithelial cells.

5. Neurons carrying impulses towards the central nervous system best describes
   a. excitatory neurons.
   b. inhibitory neurons.
   c. sensory neurons.
   d. motor neurons.
   e. interneurons.

6. Neurons carrying impulses away from the central nervous system best describes
   a. excitatory neurons.
   b. inhibitory neurons.
   c. sensory neurons.
   d. motor neurons.
   e. interneurons.

7. What is a primary function of myelin?
   a. It surrounds and protects a neuron's dendrites.
   b. It surrounds and protects a neuron's synaptic cleft.
   c. It repolarizes a neuron's membrane.
   d. It insulates the charge of a neuron from that of its neighbors.
   e. It slows the rate of nerve impulse conduction.

8. What is a primary function of myelin?
   a. It surrounds and protects a neuron's cell body.
   b. It surrounds and protects a neuron's synaptic vesicles.
   c. It speeds the rate of nerve impulse conduction.
   e. It allows the neuron's polarization to be negated by surrounding neurons.
   e. It depolarizes the neuron's membrane.

9. Which is true of a neuron at "rest?"
   a. It has a lower concentration of potassium ions inside than out.
   b. It has a higher concentration of potassium ions inside than out.
   c. It has a higher concentration of sodium ions inside than out.
   d. It has a higher concentration of chloride ions inside than out.
   e. Its inside is positively charged relative to its outside.

10. Which is true of a neuron at "rest?"
    a. It has sodium ion leak channels.
    b. It has open sodium ion gated channels.
    c. It has closed sodium ion gated channels.
    d. It has open potassium ion gated channels
    e. It has a membrane potential of approximately +70 millivolts.

11. Which is true of a neuron during depolarization?
    a. Potassium ion gated channels open.
    b. Potassium ion gated channels close.
    c. Sodium ion leak channels open.
    d. Sodium ion leak channels close.
    e. Sodium ion gated channels open.

12. Which is true of a neuron during depolarization?
    a. The charge across the membrane increases.
    b. The charge across the membrane decreases.
    c. The charge across the membrane stays the same.
    d. Sodium ions leave the neuron's interior.
    e. Potassium ions stop moving across leak channels.

13. Synaptic knobs associated with a synaptic cleft function to
    a. receive impulses from neighboring neurons.
    b. release neurotransmitter substances into the synaptic cleft.
    c. repolarize the dendrite to which it is attached.
    d. repolarize the axon to which it is attached.
    e. increase the resting potential in the dendrite to which it is attached.

14. Synaptic knobs associated with a synaptic cleft function to
    a. store neurotransmitter substances in membrane vesicles.
    b. increase the resting potential in the axon to which it is attached.
    c. insulate the cell body from neighboring neurons.
    d. depolarize the dendrite to which it is attached.
    e. depolarize the axon to which it is attached.

15. Which is a component of every reflex?
    a. interneuron
    b. neuroglial cell
    c. brain
    d. motor cortex
    e. sensory neuron

16. Which is a component of every reflex?
    a. sensory cortex
    b. cerebral cortex
    c. spinal cord
    d. motor neuron
    e. quadriceps muscle

17. In what way are the nervous and endocrine systems similar?
    a. Both use specific cellular paths to communicate with target cells.
    b. Both use chemical messengers produced by one cell and transported to a distant target cell.
    c. Both affect specific target cells.
    d. Both evoke rapid responses.
    e. Both exclusively maintain homeostasis.

18. In what way are the nervous and endocrine systems similar?
    a. Both consist of organs interconnected only by the circulatory system.
    b. Both produce chemical messengers with specific target cells immediately neighboring the cell that produced the messenger.
    c. Both maintain a -70 millivolt charge across the membranes of cells comprising the system.
    d. Both use chemical messengers.
    e. Both evoke long term changes in target cell metabolism.

19. Many known hormones are
    a. carbohydrates.
    b. phospholipids.
    c. nucleic acids.
    d. derivatives of amino acids.
    e. derivatives of carbohydrates.

20. Many known hormones are
    a. lipoproteins.
    b. steroids.
    c. derivatives of RNA.
    d. triglycerides.
    e. glycoproteins.

21. The hypothalamus
    a. acts as a link between the nervous and endocrine systems.
    b. is an important source of adrenocorticotropic hormone.
    c. responds to releasing factors produced by the pituitary gland.
    d. contains neurosecretory cells originating in the pituitary.
    e. receives blood through a system of capillaries after the blood has passed through the pituitary.

22. The posterior pituitary gland
    a. acts as a link between the nervous and endocrine systems.
    b. is an important source of adrenocorticotropic hormone.
    c. responds to releasing factors produced by the hypothalamus.
    d. contains neurosecretory cells originating in the hypothalamus.
    e. receives blood through a system of capillaries after the blood has passed through the anterior pituitary.

23. The hypothalamus monitors the condition of the body by means of
    a. sensory neurons of the somatic nervous system.
    b. blood flowing through it.
    c. motor neurons of the parasympathetic nervous system.
    d. motor neurons of the sympathetic nervous system.
    e. neurosecretory cells.

24. The hypothalamus monitors the condition of the body by means of
    a. sensory neurons of the spinal cord.
    b. blood flowing through the pituitary gland.
    c. sensory neurons.
    d. motor neurons of the central nervous system.
    e. secretory cells of the parathyroids.

25. Oxytocin
    a. stimulates milk release from the breasts.
    b. promotes milk production by the breasts.
    c. promotes increased muscle mass.
    d. stimulates release of pituitary hormones.
    e. reduces metabolic rate.

26. Growth hormone
    a. stimulates the release of milk from the breasts.
    b. promotes the production of milk by the breasts.
    c. promotes increased muscle mass.
    d. stimulates release of pituitary hormones.
    e. reduces metabolic rate.

27. The adrenal glands release glucocorticoids in response to increased blood levels of
    a. thyroid-stimulating hormone.
    b. prolactin.
    c. cortisol.
    d. adrenocorticotropic hormone.
    e. aldosterone.

28. Sex hormone and gamete production are promoted by
    a. thyroid-stimulating hormone.
    b. gonadotropic hormones.
    c. cortisol.
    d. adrenocorticotropic hormone.
    e. renin.

29. An effect of cortisol is
    a. increased loss of water through the kidneys.
    b. increased retention of sodium in the kidneys.
    c. suppression of the body's response to injury.
    d. increased mobility of stomach contents.
    e. sexual maturation.

30. An effect of aldosterone is
    a. increased loss of water through the kidneys.
    b. increased retention of sodium in the kidneys.
    c. suppression of the body's response to injury.
    d. increased mobility of stomach contents.
    e. sexual maturation.

31. The adrenal medulla secretes epinephrine in response to
    a. increased blood levels of adrenocorticotropic hormone.
    b. stimulation by sympathetic nerve fibers.
    c. stimulation by parasympathetic nerve fibers.
    d. increased blood levels of aldosterone.
    e. decreased blood levels of aldosterone.

32. Immediate effects of epinephrine secretion by the adrenal medulla include
    a. decreased heart rate.
    b. decreased metabolic rate.
    c. increased blood flow to the skin.
    d. increased contraction of intestinal muscles.
    e. increased number of cells in the blood.

33. Thyroxine and triiodothyronine are
    a. hormones of the parathyroid glands.
    b. amino acids that contain iodine.
    c. proteins that contain iodine.
    d. proteins that contain potassium.
    e. hormones of the adrenal medulla.

34. Extremely low levels of thyroxine and triiodothyronine during childhood result in
    a. acromegaly.
    b. Addison's disease.
    c. cretinism.
    d. Grave's disease.
    e. exophthalmia.

35. Insulin's primary effect is to
    a. increase heart rate.
    b. regulate sugar retention by the kidneys.
    c. regulate conversion of glycogen to glucose.
    d. regulate conversion of glucose to glycogen.
    e. stimulate cellular uptake of blood glucose.

36. Type II diabetes directly results from
    a. a failure of the pancreas to produce insulin.
    b. over production of insulin.
    c. insulin receptor abnormalities.
    d. a failure of the hypothalamus to monitor blood iodine levels.
    e. underproduction of prolactin.

## Concept Map Construction

Construct a concept map for each group of terms. Be sure to include appropriate connector phrases. You may add other terms as necessary and use terms in the plural or singular form.

1. somatic nervous system, spinal nerve, sensory receptor, parasympathetic nervous system, neurotransmitter
2. neuron, neurosecretory cell, neurotransmitter, hormone, estrogen
3. releasing factor, antidiuretic hormone, acromegaly, suppression of the immune system, protein

# Chapter 16

# Sensory Perception

## Go Figure!

Figure 16-1: Which part of a rod actually "receives" light?

Figure 16-2: Which ear bone is attached directly to the tympanic membrane?

Which ear bone is attached directly to the stirrup?

What might happen if the round window did not dissipate incoming sounds?

Figure 16-3: Based upon the definitions of tissues and organs presented in Chapter 14, how would you classify a taste bud?

What other type of sensory receptor will be stimulated by eating ice cream?

Figure 16-4: Suggest a possible reason why our sense of smell is impaired when our nasal cavity is congested.

## Matching Questions

Write the letter of the phrase that best matches the numbered term on the left. Use each only once.

_____ 1. stimulus

_____ 2. sensory receptor

_____ 3. mechanoreceptor

_____ 4. semicircular canals

_____ 5. rod cell

_____ 6. rhodopsin

_____ 7. ciliary muscles

_____ 8. accommodation

_____ 9. retina

_____ 10. hair cell

_____ 11. dark adaptation of the eye

_____ 12. olfaction

_____ 13. tympanic membrane

_____ 14. stirrup

_____ 15. basilar membrane

_____ 16. auditory nerve

_____ 17. sweet

_____ 18. taste bud

_____ 19. fovea

_____ 20. cerebral cortex

a. nonspecific category of cells capable of initiating a nervous impulse in response to a change in the environment

b. an example of a photoreceptor cell

c. a visual pigment

d. includes rod pigment activation

e. all vertebrate photoreceptor cells are found here

f. fluid-filled chambers of the vestibular apparatus

g. location of cochlea's hair cells

h. part of brain responsible for interpreting sensory input

i. contains neurons running from the cochlea to the brain

j. the act of changing the shape of the eye's lens

k. sense of smell

l. any change in the environment

m. one of the primary tastes

n. bone attached to the cochlea

o. portion of retina containing the greatest concentration of cones

p. located at the inner end of the ear canal

q. one location of chemoreceptors

r. a touch receptor, for example

s. receptor for hearing and balance

t. these change the shape of the eye's lens

## Multiple Choice Questions

1. Which is considered a sensory receptor?
   a. sclera
   b. choroid
   c. taste bud
   d. basilar membrane
   e. somatic sensory cortex

2. Which is considered a sensory receptor?
   a. cornea
   b. pupil
   c. cone cells
   d. rhodopsin
   e. tapetum

3. Mechanoreceptors respond to
   a. changes in temperature.
   b. light.
   c. chemicals that are released from damaged tissue.
   d. changes in pressure.
   e. chemicals in the air.

4. Some chemoreceptors respond to
   a. changes in temperature.
   b. light.
   c. changes in humidity.
   d. changes in pressure.
   e. chemicals in the air.

5. The outermost layer of the eye is the
   a. choroid.
   b. retina.
   c. optic nerve.
   d. sclera and cornea.
   e. iris and lens.

6. The innermost layer of the eye is the
   a. choroid.
   b. retina.
   c. optic nerve.
   d. sclera and cornea.
   e. iris and lens.

7. How does the human eye focus an image on the retina?
   a. As light passes through the sclera, light rays are focused by a change in the shape of the cornea.
   b. As light passes through the iris, light rays are focused by a change in the size of the pupil.
   c. As light passes through the lens, the rays are focused by a change in the shape of the lens.
   d. As light passes through the pupil, the rays are focused on the retina.
   e. As light strikes the iris it is reflected back to the rod and cone layer where it is focused.

8. How is the shape of the human lens changed?
   a. The fovea contracts and changes the lens' shape.
   b. Ciliary muscles contract and change the lens' shape.
   c. Ligaments attached to the lens contract and change its shape.
   d. The sclera contracts and changes the lens' shape.
   e. The muscles of the iris contract and change the lens' shape.

9. Which type of eye cell is responsible for color perception?
   a. cone
   b. ciliary muscle cell
   c. pupillary cell
   d. optic nerve cell
   e. rod

10. Which type of eye cell is responsible for black and white vision?
    a. cone
    b. ciliary muscle cell
    c. pupillary cell
    d. optic nerve cell
    e. rod

11. In the human eye, which occurs during dark adaptation?
    a. contraction of ciliary muscles
    b. inactivation of rhodopsin
    c. closure of the iris
    d. activation of cone pigments
    e. activation of rod pigments

12. In the human eye, which occurs during dark adaptation?
    a. relaxation of ciliary muscles
    b. inactivation of the optic nerve
    c. dilation of the iris
    d. inactivation of cone pigments
    e. inactivation of rod pigments

13. What is the primary type of sensory receptor in the human ear?
    a. rods
    b. cones
    c. hair cells
    d. vestibular apparatus
    e. thermoreceptor

14. What are the two primary sensory systems in the human ear?
    a. olfaction and maintenance of posture
    b. thermoreception and maintenance of posture
    c. awareness of tissue damage and thermoreception
    d. hearing and maintenance of posture
    e. hearing and maintenance of balance

15. What comprises the human middle ear?
    a. a chamber and three ear bones
    b. a chamber and the vestibular apparatus
    c. the cochlea and the vestibular apparatus
    d. the pinna and a channel leading to the tympanic membrane
    e. the vestibular apparatus and the auditory nerve

16. What comprises the inner ear in humans?
    a. a chamber and three ear bones
    b. a chamber and the vestibular apparatus
    c. the cochlea and the vestibular apparatus
    d. the ear flap and a channel leading to the tympanic membrane
    e. the vestibular apparatus and the auditory nerve

17. In the transmission of sound waves through the ear, which is affected first?
    a. cochlea
    b. semicircular canals
    c. anvil
    d. hammer
    e. tympanic membrane

18. Cochlear hair cells sit on the
    a.. vestibular apparatus.
    b. tympanic membrane.
    c. basilar membrane.
    d. tectorial membrane.
    e. stirrup.

19. Which is responsible for detecting changes in body position?
    a. vestibular apparatus
    b. cochlea
    c. middle ear
    d. external ear
    e. olfactory epithelium

20. Which IS NOT involved in transmission of sound waves to sound receptors?
    a. hammer
    b. cochlear fluid
    c. tympanic membrane
    d. semicircular canals
    e. stirrup

21. How are taste and smell receptors similar?
    a. both are found in the oral and nasal cavities
    b. both are chemoreceptors
    c. both respond to stimulation from distant sources
    d. both respond to a very few types of stimuli
    e. both are major contributors to somatic sensation

22. How are taste and smell receptors different?
    a. Taste receptors are found only in the oral cavity; smell receptors are found in the nasal cavity.
    b. Taste receptors respond only to chemical stimuli; smell receptors respond only to mechanical stimulation.
    c. Taste receptors are stimulated by direct contact with chemicals; smell receptors are stimulated by indirect contact with chemicals.
    d. Taste receptor function is well-understood; smell receptor function is not.
    e. Taste receptors are located in an epithelial layer; smell receptors are not.

## Concept Map Construction

Construct a concept map for each group of terms. Be sure to include appropriate connector phrases. You may add other terms as necessary and use terms in the plural or singular form.

1. mechanoreceptor, somatic sensory cortex, sensory neuron, somatic nervous system, retina
2. dark adaptation, visual pigment, rhodopsin, visual cortex, optic nerve
3. hypothalamus, thermoreceptor, autonomic nervous system, hair cell, vestibular apparatus

# Chapter 17

# Protection, Support, and Movement

## Go Figure!

Figure 17-1: Which layer corresponds to what is commonly called skin?

In which skin layer do sensory receptor nerve endings appear? Do you think these are part of the somatic or visceral sensory system?

Many muscles cause movement because they are attached at one end to something rigid (or something held firmly in place) and at the other to something that can move. Contraction results in movement at the mobile end. Since the muscle anchored to a hair shaft is not attached to something immobile, what is the effect of its contraction?

By what mechanism do nutrients and wastes move between cells of the dermis and living cells of the lower epidermis?

Figure 17-3: Which part of the appendage in Figure 17-3b must be rigid for contraction to produce movement?

If the upper and lower part of the appendage in Figure 17-3b were not connected with a flexible piece of exoskeleton, what would happen when the flexor muscle contracted?

What functional purpose does the presence of flexor and extensor muscles serve in this appendage?

Figure 17-4: Osteocytes remain a living part of bone even after bone elongation has stopped. But the living cells of hair and fingernails become filled with keratin and die to become part of the hair or fingernail. What functional purpose does the maintenance of living osteocytes serve in human bone?

How is this function served in hair and fingernails?

Figure 17-5: Identify at least three bones of the axial skeleton.

Identify at least three bones of the appendicular skeleton. (Note: the blue extending from each rib is costal cartilage and is considered part of the axial skeleton.)

Figure 17-6: Use Figure 17-5 and this figure to answer these questions.

To which bones is the biceps muscle attached?

Which bone remains immobile when the forearm is flexed? Which is moved?

The bone which is held immobile during this motion can be moved at other times. What keeps it immobile during flexion of the forearm?

Answer the same questions for extension of the forearm.

Figure 17-7: Why do you suppose muscle fiber nuclei are located along the outer edge of the cytoplasm rather than deep within it?

What purpose is served by having more than one nucleus in each muscle fiber?

Figure 17-8: Describe a physical reason that I bands and H zones stain lighter than the outer edges of the A band.

What happens to the distance between Z lines during contraction?

What happens to the width of the A band's darkly staining portions?

Which filament type is most abundant in a single myofibril?

Figure 17-10: The wheelbarrow handle is the lever. How does the length of the lever affect the force which must be applied?

If you lifted from a point on the handle closer to the load, would you have to apply more or less force to lift the same load?

## Matching Questions

Write the letter of the phrase that best matches the numbered term on the left. Use each only once.

| | | | |
|---|---|---|---|
| ____ 1. | hair | a. | living epithelial layer of skin |
| ____ 2. | epidermis | b. | hardened covering of arthropods |
| ____ 3. | dermis | c. | dense bone tissue |
| ____ 4. | keratin | d. | connective tissue comprising the human embryonic skeleton |
| ____ 5. | sebum | e. | connective tissue structure that holds two bones together |
| ____ 6. | hydrostatic pressure | f. | area where actin filaments are separated from myosin filaments |
| ____ 7. | exoskeleton | g. | dead, keratinized cells derived from a living follicle |
| ____ 8. | endoskeleton | h. | one contractile unit of a myofibril |
| ____ 9. | compact bone | i. | nonfatty dense connective tissue layer of skin |
| ____10. | spongy bone | j. | protein of thick filaments |
| ____11. | cartilage | k. | composed of bone and cartilage in vertebrates |
| ____12. | osteocyte | l. | primary component of hair and nails |
| ____13. | ligament | m. | controlled primarily by the motor portion of the autonomic nervous system |
| ____14. | tendon | n. | lipid mixture which coats hair and skin |
| ____15. | myosin | o. | its spaces are filled with marrow |
| ____16. | smooth muscle | p. | allows a limb to move in two directions |
| ____17. | antagonistic muscle pair | q. | connective tissue structure that joins a muscle to a bone |
| ____18. | sarcomere | r. | a bone cell |
| ____19. | I band | s. | water pressure |

## Multiple Choice Questions

1. Which is a layer of human skin?
   a. the layer of loose connective tissue below the dermis
   b. the layer of fat below the dermis
   c. the epidermis
   d. the red marrow
   e. the sarcomere

2. Which is a layer of the human integument?
   a. the dermis
   b. the sarcoplasmic reticulum
   c. the lines of compression
   d. the yellow marrow
   e. the H zone

3. Which organs are found in the human integument?
   a. blood vessels, muscles, and the basal portions of exocrine glands
   b. blood vessels, muscles, and the basal portions of endocrine glands
   c. ganglia, muscles, and the basal portions of endocrine glands
   d. ganglia, muscles, and the basal portions of hair follicles
   e. muscles and the epidermal portion of endocrine glands

4. Which tissues are found in human skin?
   a. epithelial and connective
   b. epithelial and muscle
   c. epithelial, connective, muscle, and nerve
   d. connective, muscle, and nerve
   e. muscle and nerve

5. Which DOES NOT describe an exoskeleton?
   a. In many animals, it is secreted by the epidermis.
   b. It is nonliving.
   c. It may contain joints to allow movement.
   d. It provides protection and support.
   e. Movement is produced by changing hydrostatic pressure.

6. Which DOES NOT describe a vertebrate endoskeleton?
   a. It is living.
   b. It is made of bone and/or cartilage.
   c. Movement is produced by contracting muscles.
   d. Joints are not necessary for movement.
   e. It provides protection and support.

7. What type of tissue is bone?
   a. epithelial
   b. connective
   c. muscle
   d. nerve
   e. epithelial and connective

8. What are the primary components of bone's extracellular matrix?
   a. keratin and calcium phosphate
   b. chitin and sodium phosphate
   c. collagen and calcium phosphate
   d. keratin and sodium phosphate
   e. collagen and sodium phosphate

9. The protein portion of bone
   a. gives it flexibility.
   b. gives it rigidity.
   c. provides a flexible outer covering.
   d. fills the cavity of long bones such as the femur.
   e. provides a path along which materials may diffuse into osteocytes.

10. The salt portion of bone
    a. gives it flexibility.
    b. gives it rigidity.
    c. provides a flexible outer covering.
    d. fills the cavity of long bones such as the femur.
    e. provides a path along which materials may diffuse into osteocytes.

11. Which is a bone cell?
    a. chondrocyte
    b. osteocyte
    c. epidermal cell
    d. neuron
    e. collagen

12. Red marrow is associated with
    a. spongy bone.
    b. compact bone.
    c. dermis.
    d. epidermis.
    e. cartilage.

13. Which is NOT part of the axial skeleton?
    a. sternum
    b. ribs
    c. vertebrae
    d. metatarsals
    e. cranium

14. Which is NOT part of either the pelvic or pectoral girdle?
    a. scapula
    b. clavicle
    c. carpal
    d. ilium
    e. pubis

15. The portion of a myofibril from one Z line to the next is called
    a. a muscle fiber.
    b. an H zone.
    c. a sarcomere.
    d. a sarcolemma.
    e. a thin filament.

16. The characteristic banding pattern of skeletal muscle is the consequence of overlapping
    a. myosin filaments.
    b. actin filaments.
    c. thin filaments.
    d. actin and myosin filaments.
    e. pairs of thick filaments.

17. Which describes skeletal muscles?
    a. They lack obvious striations.
    b. They have more than one nucleus per muscle cell.
    c. Their cells are more appropriately referred to as myofibrils.
    d. Their fibers are seldom anchored to bone.
    e. They are involuntary.

18. In what way are smooth and cardiac muscle similar?
    a. Both have obvious striations.
    b. Both are activated by the visceral sensory nervous system.
    c. Both are activated by the autonomic nervous system.
    d. Both act on joints of the skeleton.
    e. Both function anaerobically.

19. Which DOES NOT describe skeletal muscle contraction?
    a. Cross-bridges form between thick and thin filaments.
    b. Each myosin filament attaches only once to an actin filament for each contraction.
    c. Z-lines get closer together during a contraction.
    d. The A-band stays the same width during contraction.
    e. ATP is required.

20. Which DOES NOT describe skeletal muscle contraction?
    a. Strength of contraction is related to the number of fibers that are used.
    b. The H-zone gets smaller during contraction.
    c. The I-band gets smaller during contraction.
    d. Myosin is an enzyme that hydrolyzes the ATP necessary for contraction.
    e. Sarcomeres get larger to push the bone moved.

## Concept Map Construction

Construct a concept map for each group of terms. Be sure to include appropriate connector phrases. You may add other terms as necessary and use terms in the plural or singular form.

1. vertebrate, exoskeleton, cartilage, connective tissue, insect
2. integument, sebum, dermis, blood vessels, diffusion
3. neurotransmitter, sarcomere, skeleton, smooth muscle, mitochondrion

# Chapter 18

# Nutrition and Digestion

## Go Figure!

Figure 18-1: Which branch of the peripheral nervous system is involved in coordinating contraction of the intestines' muscular wall?

What tissue composes the luminal edge of the intestinal mucosa?

Are salivary glands exocrine or endocrine glands?

How many salivary glands are there all together? Remember this figure shows only the right side.

Identify the pharynx, epiglottis and tongue in this figure.

What role does the tongue play in swallowing?

Figure 18-2: Which branch of the peripheral nervous system carries signals from the brain to the stomach in response to the sight or smell of food?

Which branch carries signals from the stomach to the brain when food arrives at the stomach?

According to the figure, the stomach produces gastrin. What organ system would you place the stomach in based upon its production of this substance?

Figure 18-3: Where is the lumen of the small intestine in relation to the illustration on the right, towards the top or the bottom?

What type of tissue is immediately below the mucosa?

What tissue type forms the wall of a villus?

## Matching Questions

Write the letter of the phrase that best matches the numbered term on the left. Use each only once.

_____ 1. digestion
_____ 2. serosa
_____ 3. mucosa
_____ 4. mouth
_____ 5. mucin
_____ 6. saliva
_____ 7. bolus
_____ 8. pharynx
_____ 9. peristalsis
_____10. gastric juice
_____11. hydrochloric acid
_____12. chyme
_____13. alcohol
_____14. bile salts
_____15. secretin
_____16. microvilli
_____17. lacteals
_____18. sodium bicarbonate
_____19. rumen
_____20. filter feeder

a. a major protein lubricant
b. a specific nonprotein component of gastric juice
c. chemical and physical breakdown of food
d. hormone produced by endocrine cells in the wall of the small intestine
e. one of few foods absorbed in the stomach
f. glandular inner lining of the digestive tract
g. cytoplasmic extensions which act to increase the surface area of the small intestine
h. contains mucin and an enzyme for starch digestion
i. component of pancreatic juice that neutralizes acid
j. connective and epithelial tissue covering the digestive tract
k. site of cellulose digestion in cattle
l. collectively, all of the stomach's secretions
m. muscular contraction typical of gut
n. technical term for mass of swallowed food
o. obtains food suspended in water
p. back wall of the oral cavity
q. these emulsify fats
r. involved in fat absorption in the small intestine
s. food material in stomach
t. site of the initiation of starch digestion

## Multiple Choice Questions

1. Food in the human stomach next enters the
   a. pancreas.
   b. liver.
   c. esophagus.
   d. small intestine.
   e. large intestine.

2. Food in the human mouth next enters the
   a. pancreas.
   b. liver.
   c. esophagus.
   d. small intestine.
   e. large intestine.

3. Which is the longest segment of the digestive tube?
   a. esophagus
   b. stomach
   c. colon
   d. small intestine
   e. large intestine

4. Of those listed, which is the shortest segment of the digestive tube?
   a. esophagus
   b. stomach
   c. colon
   d. small intestine
   e. large intestine

5. Which IS NOT a function associated with the mouth and oral cavity?
   a. Teeth grind the food into smaller pieces.
   b. Starch is hydrolyzed.
   c. Food is mixed with a lubricating mucus to form a bolus.
   d. Lysozymes digest protein in the food.
   e. Salivary gland secretions are added in the oral cavity.

6. Which DOES NOT describe digestion in the stomach?
   a. The contents are referred to as chyme.
   b. Bile is added to the food.
   c. Hydrochloric acid is added to food.
   d. The acidic environment kills most bacteria on the food.
   e. Proteins are denatured and digested.

7. In what organ does the majority of chemical digestion occur?
   a. esophagus
   b. stomach
   c. small intestine
   d. large intestine
   e. salivary gland

8. In what organ does the majority of absorption occur?
   a. esophagus
   b. stomach
   c. small intestine
   d. large intestine
   e. salivary gland

9. Which describes chyme entering the large intestine?
   a. It has a pH of 8.2.
   b. It is very watery.
   c. It contains many nutrients which will be absorbed by the colon.
   d. It contains a great deal of pepsinogen from the stomach.
   e. It contains secretin.

10. Which describes chyme entering the large intestine?
    a. It has an acidic pH.
    b. It is very dehydrated.
    c. Most of the nutrients have already been absorbed.
    d. It contains a great deal of trypsin from the stomach.
    e. It contains cholecystokinin.

11. What do salivary glands and the pancreas have in common?
    a. Both secrete bicarbonate buffer.
    b. Both secrete carbohydrases.
    c. Both empty their secretions into the blood.
    d. Both secrete nucleases.
    e. Both secrete lipases.

12. In what way do salivary glands and the pancreas differ?
    a. Only one of them secretes carbohydrases.
    b. Only one of them is an exocrine gland.
    c. Only one of them has a tube opening directly into the digestive tract.
    d. Only one of them produces mucin.
    e. Only one of them contains epithelial tissue.

13. Which is unique to the small intestine?
    a. possession of muscle in the wall
    b. a glandular lining called mucosa
    c. possession of a serosa
    d. possession of villi
    e. susceptibility to ulceration

14. Which is unique to the small intestine?
    a. lacteals
    b. capillaries
    c. peristalsis
    d. sphincters
    e. mucus

15. Which stimulates pancreatic secretion?
    a. pepsin
    b. pepsinogen
    c. bile
    d. gastrin
    e. cholecystokinin

16. Which stimulates gastric secretion?
    a. secretin
    b. pepsinogen
    c. mucin
    d. gastrin
    e. bicarbonate ion

17. Which regulatory signal plays a role in gastric juice secretion once food has entered the stomach?
    a. somatic stimulation from the brain in response to the initial smell of food
    b. somatic stimulation of the stomach from the brain
    c. autonomic stimulation of the stomach from the brain
    d. sympathetic stimulation of the stomach from the brain
    e. the release of secretin from endocrine cells of the stomach

18. Which regulatory signal plays a role in gastric juice secretion once food has entered the stomach?
    a. visceral stimulation from the brain in response to the initial smell of food
    b. somatic sensory stimulation from the brain
    c. sympathetic motor stimulation of the stomach from the brain
    d. sympathetic sensory stimulation of the stomach from the brain
    e. the release of gastrin from endocrine cells of the stomach

19. Which is listed by your text as a product of bacterial metabolism absorbed by the human colon?
    a. vitamin $B_1$
    b. vitamin K
    c. vitamin C
    d. acetic acid
    e. pyruvate

20. Which is listed by your text as a product of bacterial metabolism absorbed by the human colon?
    a. vitamin $B_{12}$
    b. vitamin E
    c. vitamin D
    d. folic acid
    e. pyruvate

21. Which best describes filter feeders?
    a. Food particles may be trapped in mucus or strained from water.
    b. Digestion must be intracellular because the food particles are large.
    c. They possess extra stomach chambers.
    d. They are generally discontinuous feeders.
    e. They possess a cecum.

22. Which best describes ruminant animals?
    a. Food particles may be trapped in mucus or strained from water.
    b. Digestion must be intracellular because the food particles are large.
    c. They possess extra stomach chambers.
    d. They are generally continuous feeders.
    e. They possess a cecum.

## Concept Map Construction

Construct a concept map for each group of terms. Be sure to include appropriate connector phrases. You may add other terms as necessary and use terms in the plural or singular form.

1. endocrine gland, saliva, gastrin, glandular epithelium, autonomic nervous system
2. bolus, esophagus, pharynx, digestion, peristalsis
3. liver, gall bladder, enzyme, microvilli, exocrine gland

# Chapter 19

# Circulation, Excretion, and Respiration

## Go Figure!

Figure 19-1: Name two physical differences between arteries and veins illustrated by this figure.

Which basic tissue type is endothelium?

Which basic tissue type is the elastic layers?

Is there a structural difference between capillaries on the arteriole side of a capillary net and the venule side?

Figure 19-3: What do the gray arrows represent?

What do the red arrows represent?

Figure 19-4: There are valves between ventricles and atria called atrioventricular valves (on the right side the tricuspid valve, and on the left the bicuspid or mitral valve). What would happen to blood flow if these valves did not function properly?

What would happen to the blood flow pattern if blood leaked through the muscle tissue separating the right and left ventricles?

Figure 19-5: What is the direction of molecular oxygen movement in the capillary beds of a skeletal muscle?

What is the direction of molecular oxygen movement in the capillary beds of the lung?

What is the direction of carbon dioxide movement in the capillary beds of the lung?

How do arteries and veins of the pulmonary circulation differ from those of the systemic circulation?

Figure 19-8: There are three lightning bolt-shaped arrows originating at the sinoatrial node. What is the significance of their lengths relative to one another?

What do the black arrows represent?

Figure 19-9: "If red blood cells were spherical, they could hold more hemoglobin and therefore be more effective at oxygen transport." What is the logical flaw in this statement?

Figure 19-11: Your text describes blood clotting. Which aspects of that description are missing from this diagram?

At which points in the diagram do the missing factors impact the cascade of events?

Figure 19-12: Characterize the "hearts" of a grasshopper or earthworm and compare them to the human heart.

If these structures are considered functional hearts, what role must they play in circulation? What tissue must be present?

Figure 19-14: Trace the path blood travels from renal artery to renal vein. Through how many capillary systems will blood entering the glomerulus pass before reaching the renal vein?

Fluid passes from the glomerulus to Bowman's capsule. What does this suggest about the fluid pressure in each structure?

Figure 19-15: What do the black arrows illustrate?

What does the colored arrow next to the word urine have in common with the black arrows?

How do arrow color differences along tubules relate to functional differences?

What is the significance of the arrow above the word cortex?

Along which sections of the tubules are osmotic gradients responsible for the direction of the arrow?

Along which sections is pressure responsible for the direction of the arrow?

Along which sections is active transport responsible for the direction of the arrow?

Figure 19-16: Observe the vocal cord inset. In which position would air passing over the vocal cords make the lowest pitched sound?

Which position would produce the highest pitched sound?

Based upon its position, what do you think the literal meaning is for the word "epiglottis?"

Observe the lung tissue inset. Identify a single alveolus. What is the large tube at the top?

Figure 19-17: Why is blood in the pulmonary venule circulation shown in red?

Why is blood in the pulmonary arteriole circulation shown in blue?

Does blood directly touch the alveoli of a healthy lung?

How many epithelial layers separate air and blood?

How many plasma membranes (if you include the red blood cell's)?

By what mechanism do molecular oxygen and carbon dioxide move between blood and alveolus?

Figure 19-18: What does the symbol Hb stand for?

How many $O_2$ molecules can each hemoglobin bind?

Figure 19-19: In the countercurrent flow arrangement, does oxygen always move down its concentration gradient?

How does countercurrent flow affect the concentration gradient of carbon dioxide?

Figure 19-20: What tissues comprise the wall of the trachea?

What adaptive significance is there to the absence of cartilagenous rings in insect tracheae?

## Matching Questions

Write the letter of the phrase that best matches the numbered term on the left. Use each only once.

| | |
|---|---|
| ____ 1. blood pressure | a. microscopic vessel connecting artery and vein |
| ____ 2. endothelium | b. any thin, moist epithelium across which gases readily diffuse |
| ____ 3. sinoatrial node | c. specialized cardiac tissue that regulates heart contractions |
| ____ 4. capillary | d. maintenance of water and salt balance |
| ____ 5. venule | e. blood vessel intermediate in size between capillary and vein |
| ____ 6. systemic circuit | f. the epithelial lining of blood vessels |
| ____ 7. lymphatic capillary | g. functional kidney unit |
| ____ 8. antidiuretic hormone | h. subdivision of the human trachea |
| ____ 9. osmoregulation | i. its secretion increases with an increase in blood osmotic pressure |
| ____10. uric acid | j. $HCO_3^-$ |
| ____11. nephron | k. $Hb(O_2)_4$ |
| ____12. proximal convoluted tubule | l. nontoxic nitrogen compound |
| ____13. aerobic respiration | m. air pocket at the end of a bronchiole |
| ____14. respiratory surface | n. results from elastic recoil of arteries |
| ____15. cutaneous | o. kidney structure emerging from Bowman's capsule |
| ____16. glottis | p. catalyzes the formation of carbonic acid |
| ____17. bronchus | q. microscopic dead end vessels physically near blood capillaries |
| ____18. alveolus | r. location of the respiratory center |
| ____19. oxyhemoglobin | s. all vessels past the left ventricle and before the right atrium |
| ____20. bicarbonate ion | t. pertaining to the skin |
| ____21. carbonic anhydrase | u. requires an inorganic final electron acceptor |
| ____22. medulla | v. opening into the larynx |

## Multiple Choice Questions

1. Which returns blood to the human heart from the systemic circulation?
   a. arteries
   b. arterioles
   c. capillaries
   d. veins
   e. venules

2. Which returns blood to the human heart from the pulmonary circulation?
   a. arteries
   b. arterioles
   c. capillaries
   d. veins
   e. venules

3. According to your text, most veins in the human systemic circulation are transporting blood
   a. downward, with the force of gravity.
   b. upward, against the force of gravity.
   c. horizontally, neither with nor against the force of gravity.
   d. to the liver.
   e. to the kidneys.

4. What provides the majority of force behind blood movement <u>towards</u> the heart?
   a. valves
   b. gravity
   c. heart beat
   d. contraction of muscles in veins
   e. compression of veins by skeletal muscles

5. Which is most likely to occur in intestinal arterioles during times of emotional or physical stress?
   a. increased diameter
   b. decreased diameter
   c. decreased blood pressure
   d. lower resistance to blood flow
   e. total vascular shut down

6. Which is most likely to occur in skeletal muscle arterioles during times of emotional or physical stress?
   a. increased diameter
   b. decreased diameter
   c. increased blood pressure
   d. higher resistance to blood flow
   e. total vascular shut down

7. Blood in the right atrium next enters the
   a. right ventricle
   b. left atrium
   c. left ventricle
   d. aorta
   e. pulmonary artery

8. Blood in the left ventricle next enters the
   a. right ventricle
   b. left atrium
   c. left ventricle
   d. aorta
   e. pulmonary artery

9. Ventricular stimulation by the atrioventricular (AV) node results in
   a. a wave of muscular contraction beginning at the AV node and proceeding down the ventricles.
   b. a wave of muscular contraction beginning at the end of the ventricle opposite the AV node and proceeding up the ventricles.
   c. a wave of muscular contraction beginning at the AV node and proceeding down the atria.
   d. a wave of muscular contraction beginning at the end of the atrium opposite the AV node and proceeding up the atria.
   e. a wave of muscular contraction forcing blood through the bicuspid valves.

10. Atrial stimulation by the sinoatrial (SA) node results in
   a. a wave of muscular contraction beginning at the SA node and proceeding across the atria.
   b. a wave of muscular contraction beginning at the end of the atria opposite the SA node and proceeding up each atrium.
   c. a wave of muscular contraction beginning at the AV node and proceeding down the atria.
   d. a wave of muscular contraction beginning at the end of the atrium opposite the AV node and proceeding up the atria.
   e. a wave of muscular contraction forcing blood through the semilunar valves.

11. Which blood component contains clotting proteins?
   a. leukocytes
   b. white blood cells
   c. erythrocytes
   d. phagocytes
   e. plasma

12. Which forms a network of protein fibers during clot formation?
   a. hemoglobin
   b. fibrin
   c. Factor VIII
   d. lymph
   e. aldosterone

13. Which is missing from an open circulatory system?
   a. blood
   b. heart
   c. arteries
   d. veins
   e. capillaries

14. Which has an open circulatory system?
   a. human
   b. earthworm
   c. grasshopper
   d. mouse
   e. cow

15. How are the ureters and esophagus similar?
   a. Both are passive tubes through which materials move by gravity.
   b. Both are tubes leading to a passive storage organ.
   c. Both are tubes through which materials are moved by peristalsis.
   d. Both absorb water from their contents and return it to the circulatory system.
   e. Both transport metabolic wastes

16. Which is an accurate comparison of urea and uric acid?
    a. Urea is formed from the metabolic breakdown of proteins, and uric acid from catabolism of nucleic acids.
    b. Urea is formed from the metabolic breakdown of nucleic acids, and uric acid from catabolism of protein.
    c. Both are formed by the kidneys.
    d. Both are relatively nontoxic nitrogenous molecules at low concentrations.
    e. Uric acid is excreted in water, and urea as a dry paste.

17. Filtrate in the proximal convoluted tubule next enters
    a. Henle's loop.
    b. the collecting duct.
    c. the distal convoluted tubule.
    d. the glomerulus.
    e. Bowman's capsule.

18. Filtrate in Henle's loop next enters
    a. the collecting duct.
    b. the proximal convoluted tubule.
    c. the distal convoluted tubule.
    d. the glomerulus.
    e. Bowman's capsule.

19. Which DOES NOT describe ADH?
    a. It acts on the distal part of the nephron.
    b. It is released by the posterior pituitary.
    c. It causes the nephron to decrease water reabsorption.
    d. It is released when blood becomes too concentrated.
    e. Its release is under hypothalamic control.

20. Which DOES NOT describe aldosterone?
    a. It is produced by the kidney.
    b. It is released when a decrease in blood sodium is detected.
    c. It promotes sodium reabsorption by the nephron.
    d. It is a hormone.
    e. It is involved in salt homeostasis.

21. Which is a property of cellular respiratory membranes?
    a. They must be covered with a layer of moisture so that gases may dissolve in water.
    b. They are covered with keratin.
    c. They must be dry so gases will diffuse across them.
    d. They are actively endocytotic.
    e. They transport gases by facilitated diffusion.

22. Which is a property of cellular respiratory membranes?
    a. They must be covered with a layer of collagen so that gases may dissolve before diffusing across.
    b. They are covered with chitin.
    c. They must be covered with mucus so gases will diffuse across them.
    d. They are composed of thin, gas permeable epithelium.
    e. They exchange gases by active transport.

23. What is the direction of the $O_2$ concentration gradient (from highest to lowest) in an alveolus and surrounding tissues?
    a. from alveolus to bronchiole
    b. from red blood cell to alveolus
    c. from alveolus to blood plasma
    d. from blood plasma to red blood cell
    e. from erythrocyte to alveolus

24. What is the direction of the $O_2$ concentration gradient (from highest to lowest) in a systemic capillary and surrounding tissues?
    a. from tissue to blood plasma
    b. from blood plasma to red blood cell
    c. from red blood cell to blood plasma
    d. from blood plasma to erythrocyte
    e. from tissue to erythrocyte

25. Which is a correct sequence of human respiratory structures through which inhaled air passes?
    a. nasal cavity, trachea, glottis, alveoli
    b. nasal cavity, glottis, pharynx, alveoli
    c. alveoli, bronchioles, trachea, glottis, nasal cavity
    d. nasal cavity, glottis, larynx, bronchi
    e. nostrils, bronchi, larynx, nasal cavity

26. Which is a correct sequence of human respiratory structures through which exhaled air passes?
    a. nasal cavity, trachea, glottis, alveoli
    b. nasal cavity, glottis, pharynx, alveoli
    c. alveoli, bronchioles, trachea, glottis, nasal cavity
    d. nasal cavity, glottis, larynx, bronchi
    e. nostrils, bronchi, larynx, nasal cavity

Use the following information for questions 27 and 28.

Two major requirements for gas exchange include sufficient surface area across which gases move and a thin respiratory membrane. This structural arrangement allows for the homeostatic maintenance of internal $CO_2$ and $O_2$ concentrations. This large, moist surface area also presents a threat to homeostasis in terrestrial animals and aquatic animals.

27. What homeostatically maintained condition is threatened directly by the respiratory membranes of a terrestrial animal?
    a. water content of the body
    b. blood glucose level
    c. vasodilation
    d. blood calcium ion level
    e. neuron resting potential

28. What homeostatically maintained condition is threatened directly by the respiratory membranes of an aquatic animal living in a hypotonic environment?
    a. trypsin level in the stomach
    b. liver glycogen content
    c. sodium content of tissues
    d. carbonic anhydrase concentration
    e. oxygen concentration in the alveoli

29. Which is a function of the nasal passages?
    a. They filter air before it enters the bronchi.
    b. The regulate the volume of air reaching the lungs.
    c. They contract to force air into the lungs.
    d. They expand to draw air out of the lungs.
    e. They provide cartilaginous support for the bronchioles.

30. Which is a function of the nasal passages?
    a. They vibrate when air is forced over them to amplify sounds produced by the vocal cords.
    b. They moisten the air before it enters the bronchi.
    c. They add mucin to dry the air before it enters bronchioles.
    d. They prevent food or drink from passing into the glottis.
    e. The provide cartilagenous support for the trachea.

31. Which blood component transports the largest quantity of oxygen?
    a. plasma
    b. red blood cell
    c. platelet
    d. carbonic anhydrase
    e. carbaminohemoglobin

32. Which blood component transports the largest quantity of carbon dioxide?
    a. plasma
    b. red blood cell
    c. platelet
    d. carbonic anhydrase
    e. carbaminohemoglobin

33. The lamellae of gill filaments serve to
    a. increase surface area for gas exchange.
    b. direct blood flow perpendicular to the flow of water over the gills.
    c. cover and protect delicate gill membranes.
    d. channel blood flow in the same direction as water flow over the gills.
    e. establish a large chamber where oxygen-poor blood can pool.

34. Each insect spiracle leads directly into
    a. a trachea.
    b. a tracheole.
    c. a muscular diaphragm.
    d. a bronchiole.
    e. a lamella.

## Concept Map Construction

Construct a concept map for each group of terms. Be sure to include appropriate connector phrases. You may add other terms as necessary and use terms in the plural or singular form.

1. peripheral nervous system, smooth muscle, capillary, lactic acid, molecular oxygen
2. hypothalamus, ADH, aldosterone, collecting duct, glomerulus
3. red blood cell, hemoglobin, lactic acid, electron transport chain, endothelium

# Chapter 20

# Internal Defense: The Immune System

## Go Figure!

Figure 20-1: What general type of cellular transport are these cells involved in?

Once inside, what organelle provides enzymes to destroy the bacteria?

Figure 20-2: Leukocytes are able to squeeze between endothelial cells. What else will be able to leave the capillary through these spaces?

Figure 20-3: In addition to an immune system role, what other functions are performed by bone marrow?...the spleen?...lymphatic tissue?

Figure 20-4: What protein is released by killer T cells?

Killer T cells destroy tumor cells, yet, people still get cancer. How?

Figure 20-5: Where are the constant (C) regions of each chain?

Where are the variable (V) regions of each chain?

How many antibody binding sites are illustrated?

If an antibody is able to bind an antigen, what, aside from shape, must be complementary between immunoglobulin and antigen?

How many polypeptides compose one heavy chain?

Based upon what you have learned about molecular genetics, what does this imply about the number of genes coding for one heavy chain?

If there are many different heavy chains, all different in the variable region, what does this imply about the number of genes coding for heavy proteins of class G antibodies?

What does it imply about the nucleotide sequence of these genes?

Figure 20-6: The gene shown coding for one heavy chain has a section of nucleotides which are transcribed, but not translated. Is this unique to antibody genes?

What might happen if these regions were translated?

What is the cellular site of transcription?...of translation?

Figure 20-7: What is the relationship of stem cell and lymphocyte?

What is the relationship of lymphocyte and macrophage?

What is the relationship of lymphocyte and plasma cell?

## Matching Questions

Write the letter of the phrase that best matches the numbered term on the left. Use each only once.

____ 1. pus
____ 2. interferon
____ 3. macrophage
____ 4. complement
____ 5. B lymphocyte
____ 6. cytotoxic T cell
____ 7. helper T cell
____ 8. suppressor T cell
____ 9. perforins
____10. interleukin II
____11. histocompatibility antigen
____12. tissue graft
____13. cyclosporine
____14. immunoglobulin
____15. antigen binding site
____16. clonal selection
____17. clonal deletion
____18. hybridoma
____19. passive immunity
____20. immunological recall

a. formed from a large number of leukocytes
b. temporary condition produced with injection of antibodies
c. model describing immune tolerance
d. specific cell-surface antigens
e. antigen stimulating B cells that produce the antibody against that antigen
f. may produce plasma cells
g. an antibody
h. produces monoclonal antibodies
i. cell that inhibits lymphocyte activity
j. main cell infected by HIV
k. protein produced by host cells in response to a viral infection
l. released by helper T lymphocytes
m. fungal compound that suppresses cell-mediated immunity
n. cell that destroys cancer cells
o. immune system capability made possible by memory cells
p. a group of plasma proteins which act nonspecifically
q. portion of an antibody complementary to a foreign protein or polysaccharide
r. proteins released by cytotoxic T lymphocytes
s. the surgical movement of tissue from one area or one individual to another
t. derived from monocytes

## Multiple Choice Questions

1. Which is a component of the human body's nonspecific immune response?
   a. complement
   b. T cells
   c. B cells
   d. immunoglobulins
   e. plasma cells

2. Which is a component of the human body's specific immune response?
   a. fever
   b. neutrophils
   c. T cells
   d. complement
   e. interferon

3. Which is the source of antibodies in body fluids?
   a. natural killer cells
   b. macrophages
   c. B cells
   d. immunoglobulins
   e. plasma cells

4. Which group of molecules contains a protein capable of binding to a bacterium and putting holes in its plasma membrane?
   a. immunoglobulins
   b. complement
   c. interferon
   d. antibodies
   e. interleukin II

5. Inflammation is initiated
   a. when leukocytes release complement.
   b. when neutrophils release interferon.
   c. when lymphocytes release chemicals into interstitial fluid.
   d. when damaged tissue cells release chemicals into interstitial fluid.
   e. when killer T cells are damaged.

6. Inflammation includes
   a. interferon binding to a bacterium and putting holes in its plasma membrane.
   b. dilation of local blood vessels.
   c. activation of local tissue cells by interleukin II.
   d. inhibition of vasoconstriction by suppressor T cells.
   e. production of antibodies.

7. Which produces interferon?
   a. natural killer cells
   b. macrophages
   c. B cells
   d. tissue cells
   e. plasma cells

8. Interferon
   a. binds to a bacterium and puts holes in its plasma membrane.
   b. attracts phagocytic leukocytes including neutrophils.
   c. blocks a cell's ability to produce viral proteins.
   d. initiates an immune response.
   e. interferes with suppressor T cell functions.

9. Plasma cell antibody formation requires the presence of
   a. complement.
   b. antigens.
   c. basophils.
   d. neutrophils.
   e. eosinophils.

10. Macrophages are produced from
   a. B cells.
   b. monocytes.
   c. Killer T cells.
   d. neutrophils.
   e. basophils.

11. How do B cells and plasma cells compare?
   a. B cells are derived from eosinophils, while plasma cells develop from macrophages.
   b. B cells have more rough endoplasmic reticulum than plasma cells.
   c. Both B cells and plasma cells produce perforins.
   d. Both are phagocytic.
   e. B cells convert into plasma cells.

12. How do B cells and plasma cells compare?
   a. B cells are derived from macrophages, while plasma cells develop from eosinophils.
   b. B cells have less rough endoplasmic reticulum than plasma cells.
   c. B cells produce perforins, while plasma cells do not.
   d. Both are formed within the thymus.
   e. B cells produce antibodies, while plasma cells do not.

13. How do B cells and T cells compare?
   a. B cells are leukocytes, while T cells are lymphocytes.
   b. B cells mature into T cells.
   c. Both are lymphocytes.
   d. Both produce perforins.
   e. B cells are lymphocytes, while T cells are leukocytes.

14. How do B cells and T cells compare?
   a. T cells become plasma cells, while B cells become memory cells.
   b. T cells mature into B cells.
   c. Both are eosinophils.
   d. Both are involved in cell-mediated immunity.
   e. B cells are involved in soluble immune responses, while T cells participate in cell-mediated immunity.

15. Which is a function of T cells? T cells may participate
   a. in antigen production.
   b. in perforin production.
   c. in interferon production.
   d. in complement production.
   e. directly in immunoglobulin production.

16. Which is a function of T cells? T cells may
   a. participate directly in antibody production
   b. undergo mitosis to produce plasma cell clones
   c. participate in histocompatibility antigen production
   d. participate in heavy chain Ig production
   e. participate in interleukin II production

17. Commonly, human antibodies (IgG) are composed of ______ protein molecules?
   a. 1
   b. 2
   c. 3
   d. 4
   e. 5

18. The IgG class of human antibodies typically has _____ heavy chains and ______ light chains.
   a. 1, 1
   b. 2, 2
   c. 1, 2
   d. 2, 1
   e. 3, 3

19. What is unusual about the "gene" coding for the heavy chains of a human Ig antibody?
   a. It actually contains the nucleotide sequence for both heavy and light chains.
   b. It contains the nucleotide sequence to make several different light chains.
   c. It is actually two separate genes.
   d. It runs the length of one whole chromosome.
   e. It is always inherited from the mother.

20. What is unusual about the "gene" coding for the light chains of a human Ig antibody?
   a. It actually contains the nucleotide sequence for both heavy and light chains.
   b. It contains the nucleotide sequence to make several different light chains.
   c. It is actually two separate genes.
   d. It runs the length of one whole chromosome.
   e. It is always inherited from the mother.

## Concept Map Construction

Construct a concept map for each group of terms. Be sure to include appropriate connector phrases. You may add other terms as necessary and use terms in the plural or singular form.

1. immune system, circulatory system, digestive system, lymphatic system, skeletal system
2. leukocyte, lymphocyte, phagocytic cell, thymus, platelet
3. transcription, intron, variable region, constant region, protein

# Chapter 21

# Reproduction and Development

## Go Figure!

Figure 21-1: What kind of cell division is involved in asexual reproduction in multicellular animals?

Figure 21-2: What structure passes through the middle of the corpus spongiosum?

Beginning with the site of their production, name the tubes through which sperm pass on their way out of the body.

Name in sequence the accessory glands sperm pass by on their way out of the body.

Figure 21-3: On the scanning electron micrograph, identify spermatogonia, primary spermatocytes, secondary spermatocytes, spermatids, and immature sperm. State if each is haploid or diploid.

Figure 21-4: Describe the path followed by sperm in reaching the typical point of fertilization in a human female.

Describe the path followed by a baby during a noncaesarean birth.

Based upon this drawing, why do you think so many women experience frequent urination late in a pregnancy?

Figure 21-5: What do the arrows between stages represent?

Identify the potential ovum in each preovulation stage.

Figure 21-6: Give the ploidy of each cell type shown (oogonium [plural = oogonia], oocyte and polar body).

Figure 21-7: What do the relative lengths of the blue arrows indicate?

What do their relative widths indicate?

What are the relative levels of FSH, LH, estrogen, and progesterone on Day 1 of the menstrual cycle?...Day 7?...Day 14?...Day 21?...Day 28?

Figure 21-8: On the micrograph, identify the acrosome and flagellum.

What role does the acrosome play in fertilization?

How many chromosomes are in a normal human sperm?

Figure 21-9: Are the cells haploid or diploid?

Is the furrow forming in the photomicrograph (top row, middle column) a mitotic event?

What specific cellular component causes this furrow?

Figure 21-10: Are different size blastomeres distributed symmetrically in an embryo?

How does the position of the largest sea urchin blastula cells compare to the position of the largest frog blastula cells?

How does the position of yolk in each blastula compare?

Figure 21-11: What does the yellow-shaded area represent?

How many germ layers are present in the blastula?

How many germ layers are present in the early gastrula?

How many germ layers are present in the late gastrula?

What does "meso" in the term mesoderm mean? How is this an appropriate name?

Figure 21-13: Might a block of mesoderm tissue from a very early embryo develop normally for the recipient of the graft?

If so, what must happen at the molecular level?

Figure 21-15: Where does fertilization take place?

Where does cleavage begin?

To which germ layer does the trophoblast correspond?

From which group of cells will all three germ layers arise?

What do the green arrows indicate in the upper part of this drawing?

What purpose do the green arrows serve in the bottom four drawings?

## Matching Questions

Write the letter of the phrase that best matches the numbered term on the left. Use each only once.

____ 1. asexual reproduction

____ 2. gonads

____ 3. fertilization

____ 4. spermatids

____ 5. interstitial cells

____ 6. epididymis

____ 7. acrosome

____ 8. ovum

____ 9. gonadotropins

____10. endometrium

____11. oogonium

____12. larva

____13. blastomere

____14. homeotic gene

____15. egg activation

____16. germ layer

____17. ectoderm

____18. organogenesis

____19. archenteron

____20. dorsal lip

____21. villi

____22. extraembryonic membrane

____23. amniotic fluid

a. space which will give rise to the digestive system lumen

b. affects the spatial arrangement of the body parts

c. male haploid cells

d. general term referring to reproduction in the absence of fertilization

e. examples include the chorion and allantois

f. cell mitotically derived from the zygote

g. germ layer which forms the nervous system

h. inner lining of the uterus

i. self-feeding immature form

j. found between the seminiferous tubules

k. sperm component containing digestive enzymes

l. premeiotic germ cell in a female

m. the primary reproductive organs

n. marked by an increase in electrical activity

o. one egg

p. fusion of two different cell types during reproduction

q. hormones with gonads as targets

r. sits atop a testis

s. branching projections from the chorion

t. follows neurulation

u. suspends and cushions the developing mammalian embryo

v. a fold of tissue above the blastopore

w. includes the mesoderm

## Multiple Choice Questions

1. Which describes asexual reproduction?
   a. It introduces great variability to a population by mutation.
   b. It avoids "genetic monotony".
   c. It does not involve fertilization.
   d. It involves fertilization
   e. It involves the fusion of two diploid cells.

2. Which describes sexual reproduction?
   a. It introduces great variability to a population by mutation.
   b. It may involve budding of haploid cells.
   c. It does not involve fertilization.
   d. It involves fertilization
   e. It involves the fusion of two diploid cells.

3. Asexual reproduction differs from sexual reproduction in that
   a. asexual reproduction involves the fusion of two cell types, sexual reproduction does not.
   b. asexual reproduction does not require that a diploid organism undergo meiosis to maintain the diploid number of chromosomes in the next generation, sexual reproduction does.
   c. sexual reproduction produces clones of the parent, asexual reproduction does not.
   d. asexual reproduction requires the production of gametes, sexual reproduction does not.
   e. fission is a form of asexual reproduction, while budding is a form of sexual reproduction.

4. Asexual and sexual reproduction are similar in that
   a. both involve gametes.
   b. both produce variability in the offspring by creating new chromosomes (genetic recombination).
   c. both result in the passing of chromosomes from one generation to the next.
   d. both require the production of haploid cells.
   e. both require the production of diploid cells.

5. The human penis is an organ which
   a. provides a means of achieving fertilization in a terrestrial environment.
   b. provides a storage site for the maturation of sperm.
   c. produces sperm.
   d. contains four cylinders of erectile tissue.
   e. becomes erect when stimulated by sympathetic motor neurons.

6. The human penis is an organ which
   a. becomes erect when parasympathetic motor neurons stimulate the striated muscle lining its arterioles.
   b. provides an avenue of urine excretion from the body.
   c. produces the sex hormone testosterone.
   d. is the only structure of the body without parasympathetic nerve innervation.
   e. becomes erect in response to release of FSH from the anterior pituitary.

7. What would be the consequence of the corpus spongiosum filling with blood?
   a. The penis would become rigid.
   b. The penis would become flaccid.
   c. Semen would be ejaculated.
   d. Semen would be unable to pass through the urethra.
   e. Parasympathetic motor neurons would cause vasoconstriction of arterioles leading to the corpus spongiosum.

8. What would be the consequence of the arterioles in the corpus spongiosum constricting?
   a. The penis would become rigid.
   b. The penis would become flaccid.
   c. Semen would be ejaculated.
   d. Semen would be unable to pass through the urethra.
   e. Parasympathetic motor neurons would cause vasoconstriction of arterioles leading to the corpus spongiosum.

9. What is produced by seminiferous tubules?
   a. sperm
   b. testosterone
   c. FSH
   d. digestive enzymes
   e. ejaculatory fluid

10. What is produced by interstitial cells of the testes?
   a. sperm
   b. testosterone
   c. FSH
   d. digestive enzymes
   e. ejaculatory fluid

11. Which cell produces primary spermatocytes?
   a. spermatogonia
   b. secondary spermatocytes
   c. spermatids
   d. sperm
   e. interstitial cells

12. Which cell produces spermatogonia?
   a. spermatogonia
   b. primary spermatocytes
   c. spermatids
   d. sperm
   e. interstitial cells

13. Which is a function of ejaculatory fluid?
   a. It inhibits the hypothalamus from producing and releasing GnRH.
   b. It inhibits the hypothalamus from producing and releasing FSH.
   c. It inhibits the anterior pituitary from releasing LH.
   d. It temporarily neutralizes vaginal acidity.
   e. It temporarily neutralizes the effects of LH.

14. Which is true of ejaculatory fluid?
   a. It stimulates the hypothalamus to produce and release GnRH.
   b. It stimulates the hypothalamus to produce and release FSH.
   c. It provides an energy source for sperm motility.
   d. It temporarily neutralizes vaginal alkalinity.
   e. It comprises about ten percent of the material released during ejaculation.

15. Which is a function of oviducts?
   a. They are the site of zygote implantation.
   b. They are the site of oogonia development and maturation.
   c. They are they normal site of fertilization.
   d. They are the birth canal.
   e. They are the site of follicle development.

16. Which is a function of the follicle?
    a. It produces testosterone.
    b. It produces estrogen.
    c. It produces FSH.
    d. It is the normal site of fertilization.
    e. It is the location of the endometrium.

17. Oogonia divide to produce
    a. oocytes.
    b. the first polar body.
    c. the second polar body.
    d. cells of the corpus luteum.
    e. cells of the follicle.

18. How many oogonia is a human female infant born with?
    a. 0
    b. 400
    c. 100,000
    d. 500,000
    e. more than 1 million

19. What is the condition of the endometrium on Day 1 of the menstrual cycle?
    a. It is growing thicker.
    b. It is breaking down.
    c. It is beginning to secrete estrogen.
    d. It is beginning to secrete progesterone.
    e. It is inhibiting production and release of FSH.

20. What is the condition of the follicle on day 14 of the menstrual cycle?
    a. It is beginning to increase in size.
    b. It is beginning to decrease in size.
    c. It has, or is about to rupture and release a maturing oocyte.
    d. It is beginning to secrete estrogen.
    e. It is stimulating the production and release of FSH.

21. Which is the most effective contraceptive method?
    a. tubal ligation
    b. female birth control pills
    c. IUD
    d. diaphragm used with a spermicide
    e. sponge

22. Which is the most effective contraceptive method?
    a. male condom
    b. condom used with a spermicide
    c. rhythm method
    d. vasectomy
    e. rhythm method

23. Most animal embryos contain a store of food called yolk. It includes
    a. monosaccharides.
    b. disaccharides.
    c. polysaccharides.
    d. dipeptides.
    e. glycerol.

24. Most animal embryos contain a store of food called yolk. It includes
    a. cellulose.
    b. fatty acids.
    c. tripeptides.
    d. protein.
    e. nucleic acids.

25. Which is a primary sperm function?
    a. form an extracellular membrane around the egg
    b. reduce the cytoplasmic content of the egg
    c. introduce $Ca^{++}$ ions into the egg
    d. contribute a homologous chromosome set
    e. initiate meiosis in the egg

26. Which is a primary sperm function?
    a. form an intracellular membrane in the egg
    b. increase egg's cytoplasmic content
    c. reduce $Ca^{++}$ ion concentration in the egg
    d. reduce the egg's mRNA content
    e. egg activation

27. How does the size of a 16-celled sea urchin embryo compare to that of the zygote?
    a. It is approximately 16 times the zygote's size.
    b. It is approximately 8 times the zygote's size.
    c. It is approximately the same size as the zygote.
    d. It is approximately 1/2 the zygote's size.
    e. It is approximately 1/16 the zygote's size.

28. How does the size of a 16-celled frog embryo compare to that of the zygote?
    a. It is approximately 16 times the zygote's size.
    b. It is approximately 8 times the zygote's size.
    c. It is approximately the same size as the zygote.
    d. It is approximately 1/2 the zygote's size.
    e. It is approximately 1/16 the zygote's size.

29. If the amount of yolk is the primary determinant of blastocoel size, which animal has the smallest blastocoel?
    a. sea urchin
    b. human
    c. frog
    d. whale
    e. snake

30. If the amount of yolk is the primary determinant of blastocoel size, which animal has the largest blastocoel?
    a. lizard
    b. salamander
    c. quail
    d. sea urchin
    e. hawk

31. Which is formed from endoderm?
    a. spinal cord
    b. skeleton
    c. middle ear
    d. thyroid gland
    e. muscles

32. Which is formed from ectoderm?
    a. brain
    b. lungs
    c. gonads
    d. dermis
    e. blood vessels

33. Which appears first during human embryo development?
    a. blastocyst
    b. amniotic cavity
    c. chorion
    d. chorionic villi
    e. placenta

34. At what stage does implantation occur in humans?
    a. zygote
    b. blastocyst
    c. gastrula
    d. embryo
    e. fetus

## Concept Map Construction

Construct a concept map for each group of terms. Be sure to include appropriate connector phrases. You may add other terms as necessary and use terms in the plural or singular form.

1. birth control pill, FSH, follicle, oocyte, mitosis
2. mitosis, transcription, blastomere, blastopore, mesoderm
3. nervous system, ectoderm, embryonic disk, chorion, yolk sac

# Chapter 22

# Mechanisms of Evolution

## Go Figure!

Figure 22-1: The evidence supplied about the mudskipper is used to support evolution from an aquatic ancestor. Can you supply alternative explanations that are also consistent with the evidence?

Why do you think biologists favor the evolutionary model?

Figure 22-3: Which kind of selection is illustrated by the peppered moth population?

Why hadn't the black phenotype disappeared completely prior to the Industrial Revolution in England?

Figure 22-4: Which type(s) of selection might lead to formation of new species?

Figure 22-5: Under what environmental conditions have these "living fossils" lived?

Figure 22-7: Why are displays such as the ones shown in the figure, or others, such as courtship rituals, characteristic of males?

Figure 22-8: Which human characteristics are examples of sexual dimorphism?

Figure 22-9: Why do zoos send animals on "breeding loan?"

Figure 22-10: Why do you suppose African and Indian elephants are placed in different species?

Is the criterion used for labeling animal populations as different species applicable to members of all kingdoms?

Figure 22-11: Under what conditions may divergent evolution occur?

Which type of selection is likely to be involved in divergent evolution?

Besides selection, which other mechanisms may be involved?

Under what conditions may adaptive radiation occur?

Which type of selection is likely to be involved in adaptive radiation?

Under what conditions may convergent evolution occur?

How may a graph of selective pressure (like the ones in Figure 22-4) be modified to illustrate convergent evolution?

Under what conditions may parallel evolution occur?

Which type of selection is likely to be involved for each species?

How may a graph of selective pressure (like the ones in Figure 22-4) be modified to illustrate parallel evolution?

Figure 22-12: Animals as diverse as giraffes, mice, bats and gorillas are classified as members of the same class (Mammalia). Why aren't dolphins, seals, penguins and ichthyosaurs at least placed in the same class since they look more alike than giraffes, mice, bats and gorillas?

Figure 22-14: What does the horizontal axis of these graphs represent?

What do the horizontal branch points indicate about the punctuated equilibrium model?

Why does the gradualism model have acute angles at the branch points?

## Matching Questions

Write the letter of the phrase that best matches the numbered term on the left. Use each only once.

____ 1. evolution

____ 2. population

____ 3. gene pool

____ 4. gene flow

____ 5. genetic drift

____ 6. mutation

____ 7. environment

____ 8. stabilizing selection

____ 9. directional selection

____10. disruptive selection

____11. sexual selection

____12. allopatric speciation

____13. sympatric speciation

____14. postzygotic isolating mechanism

____15. reproductive isolation

____16. divergent evolution

____17. coevolution

____18. extinction

____19. gradualism

____20. punctuated equilibrium

a. most often seen in an unchanging environment

b. all of the alleles in a population

c. an evolutionary model in which changes occur in small steps

d. random changes in allele frequency

e. an essential step in speciation because it prevents exchange of alleles between populations

f. production of new species in different geographical regions

g. hybrid inviability is an example

h. a form of nonrandom mating

i. physical, physiological or behavioral change in a population over time

j. intermediate phenotypes are at a disadvantage

k. pattern of evolution in which two populations act as selective agents on each other

l. the biological unit that evolves

m. polyploidy is an example

n. the agent of natural selection

o. immigration and emigration are examples

p. pattern of evolution in which two species are formed from a single ancestral type

q. one extreme phenotype is favored, but the other is not

r. loss of a species

s. a model of evolution in which speciation occurs in relatively short periods of time

t. the mechanism by which new alleles arise

## Multiple Choice Questions

1. All of the alleles in a population are called the
   a. gene pool.
   b. genetic drift.
   c. allele frequency.
   d. species.
   e. polypeptides.

2. How common an allele is in a population is called the
   a. gene pool.
   b. genetic variation.
   c. allele frequency.
   d. dominant allele.
   e. recessive allele.

*NOTE: Answers for questions 3 and 4 are provided in the Appendix. Questions 5 and 6 constitute a similar sequence, but the answers are not given.*

3. A certain gene has two alleles. Twenty-five out of every 100 individuals are heterozygous for this gene, whereas 5 out of every 100 are homozygous recessive. What is the frequency of the recessive allele in a population of 100,000 individuals?
   a. 3%
   b. 30%
   c. 35%
   d. 17.5%
   e. 1.75%

4. What is the frequency of the dominant allele in question #3?
   a. 82.5%
   b. 98.25%
   c. 97%
   d. 70%
   e. 65%

5. There is another gene in the population that has two alleles. Twenty-five out of every 75 individuals are heterozygous for this gene, whereas 15 out of every 150 individuals are homozygous recessive. What is the frequency of the recessive allele in a population of 1000 individuals?
   a. 10%
   b. 26.5%
   c. 33.3%
   d. 40%
   e. 50%

6. What is the frequency of the dominant allele in question #5?
   a. 50%
   b. 60%
   c. 66.7%
   d. 73.5%
   e. 90%

7. Which is a requirement for a population to reach genetic (Hardy-Weinberg) equilibrium?
   a. The population must be very large.
   b. The population must not mate.
   c. All immigration into the population must be balanced by emigration out of the population.
   d. Stabilizing selection is the only kind of selection that can occur.
   e. All mutations must produce beneficial alleles.

8. Which is a requirement for a population to reach genetic (Hardy-Weinberg) equilibrium?
   a. Sexual selection is the only kind of selection that can occur.
   b. Diversifying selection is the only kind of selection that can occur.
   c. Directional selection is the only kind of selection that can occur.
   d. The population may be small.
   e. Mutation must not occur.

9. Which is an example of gene flow?
   a. migration
   b. nonrandom mating
   c. sexual selection
   d. chance fluctuations in allele frequencies
   e. polyploidy

10. Which is an example of genetic drift?
   a. migration
   b. nonrandom mating
   c. sexual selection
   d. chance fluctuations in allele frequencies
   e. polyploidy

11. Which describes stabilizing selection?
   a. It will result in speciation if given enough time.
   b. Extreme phenotypes are at an advantage.
   c. The least common phenotype is at an advantage.
   d. It is most often observed in an unchanging environment.
   e. The most common phenotype is at a disadvantage.

12. Which describes disruptive selection?
   a. It will not produce change in the population.
   b. It is common in populations with a large geographic range.
   c. The average phenotype is at an advantage.
   d. Both extreme phenotypes are at a disadvantage.
   e. Change in the population occurs in a single direction.

13. One species gradually evolving into another species best describes
   a. adaptive radiation.
   b. phyletic speciation.
   c. divergent evolution.
   d. convergent evolution.
   e. hybridization.

14. Which DOES NOT describe sympatric speciation?
   a. reproductive isolation is achieved
   b. it is an example of divergent evolution
   c. it may occur by hybridization
   d. the populations occupy different geographic regions
   e. it appears to be more common in plants than animals

15. Two closely related species evolving similar adaptations over a long period of time describes
    a. convergent evolution.
    b. adaptive radiation.
    c. parallel evolution.
    d. coevolution.
    e. extinction.

16. Species evolving as a result of selective pressures exerted on each other describes
    a. convergent evolution.
    b. adaptive radiation.
    c. parallel evolution.
    d. coevolution.
    e. extinction.

17. *Festuca viridula* and *Festuca reflexa* belong to the grass family (Gramineae). *F.viridula* is found in subalpine meadows above 6000 feet and flowers from July to August. *F. reflexa* grows in dry, open places below 5000 feet, and flowers from April to June. Based upon the information given, which describes these fescues?
    a. They are reproductively isolated by a prezygotic isolating mechanism.
    b. They are not geographically isolated.
    c. They are the products of stabilizing selection.
    d. They are the products of convergent evolution.
    e. They are the products of allopatric speciation.

18. *Lupinus Tidestromii* and *Lupinus gracilentus* belong to the legume family of angiosperms. *L. Tidestromii* is found in the coastal strand habitat and produces flowers from May to June, whereas *L. gracilentus* is found in subalpine habitats above 8000 feet and produces flowers from July to August. Based upon the information given, which describes these lupines?
    a. They are reproductively isolated by postzygotic isolating mechanism.
    b. They are geographically isolated.
    c. They are the products of convergent evolution.
    d. They are the products of sympatric speciation.
    e. They are the products of parapatric speciation.

19. The species of bromegrass (genus *Bromus*) exhibit a variety of chromosome numbers. *B. ciliatus, B. pseudolaevipes, B. vulgaris* (2n = 14), *B. mollis, B. Richardsonii, B. mollis* (2n = 28), *B. catharticus, B. arizonicus, B. marginatus* (2n = 42), and *B. inermis, B. breviaristatus, B. carinatus, B. erectus* (2n = 56) are several examples. Which is consistent with the information given?
    a. The bromegrasses provide an example of polymorphism.
    b. The bromegrasses provide an example of polyploidy.
    c. The populations had to be geographically isolated prior to speciation.
    d. The number of chromosomes in the common ancestor was probably 2n = 7.
    e. The number of chromosomes in the common ancestor was probably 2n = 28.

20. Bluegrasses belong to the genus *Poa*. Species of *Poa* have a variety of chromosome numbers, including *P. annua*, *P. Bolanderi*, *P. rhizomata* (2n = 28), *P. confinis*, *P. compressa, P. Hanseni* (2n = 42), *P. Kelloggii* (2n = 56), and *P. epilis* (2n = 56 or 84). Which is consistent with the information given?
    a. They provide an example of polymorphism.
    b. They provide an example of convergence.
    c. The number of chromosomes in the common ancestor was probably 2n = 14.
    d. The number of chromosomes in the common ancestor was probably 2n = 28.
    e. They provide an example of genetic drift.

21. The theory of punctuated equilibrium is best supported by a fossil record with
    a. morphological forms disappearing, then reappearing.
    b. long periods where the fossils show little change separated by short periods where there is a lot of change.
    c. numerous intermediate forms which suggest a gradual change.
    d. numerous different morphological forms in each period.
    e. very little change in the fossils.

22. Phyletic evolution is best supported by a fossil record with
    a. numerous gaps between morphological forms.
    b. long periods where the fossils show little change separated by short periods where there is a lot of change.
    c. numerous intermediate forms which suggest a gradual change.
    d. numerous different morphological forms in each period.
    e. very little change in the fossils.

Use the phylogenetic tree below for questions 23 and 24.

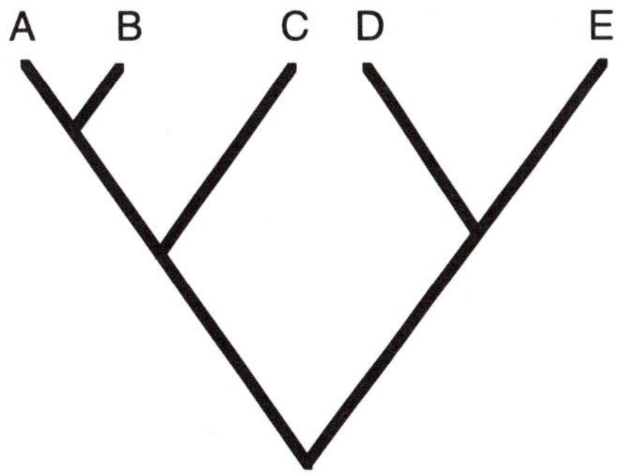


23. Which is consistent with the phylogenetic tree?
    a. Species A and B are the most closely related.
    b. Species C and D are the most closely related.
    c. Species D is more closely related to Species C than to Species E.
    d. If Species A and C belong to one genus, then Species B belongs to another genus.
    e. Species A and B are the products of directional selection.

24. Which is consistent with the phylogenetic tree?
    a. Species A is more closely related to Species C than to Species B.
    b. If Species B and D belong to one genus, then Species C belongs to another genus.
    c. Species A, B and C are the products of adaptive radiation.
    d. Species D and E are the products of convergent evolution.
    e. Species D and E are the products of disruptive selection.

## Concept Map Construction

Construct a concept map for each group of terms. Be sure to include appropriate connector phrases. You may add other terms as necessary and use terms in the singular or plural form.

1. disruptive selection, population, gene flow, Hardy-Weinberg Equilibrium, allele frequency
2. sympatric speciation, reproductive isolation, gene pool, parallel evolution, dominant allele
3. punctuated equilibrium, extinction, sexual selection, meiosis, DNA

# Chapter 23

# Evidence for Evolution

## Go Figure!

Figure 23-1: Since the cycad and the true palm are not closely related, are their palm-like characteristics homologous or analogous?

Are the leaves themselves homologous or analogous?

Figure 23-2: Why do you suppose the forelimb skeletons are not compared to a primitive vertebrate such as a fish?

Figure 23-4: What reptilian features did *Archaeopterix* have?

What avian (bird) features did *Archaeopterix* have?

If only skeletal remains had been found, how would *Archaeopterix* probably have been classified - bird or reptile?

Figure 23-5: Jean Baptiste de Lamarck proposed a theory of evolution prior to Charles Darwin. The theory involved inheritance of acquired characteristics. That is, modifications accumulated by an organism throughout its life could be passed on to its offspring. Lamarck might modify this proposal to account for vestigial structures by saying organs not used by an organism during its life are passed on to offspring in some atrophied form. How might Darwin respond to this proposal?

Figure 23-6: What are the major differences between the skulls of *Australopithecus africanus* and *Homo sapiens*?

Which skull does *Homo erectus* most closely resemble? Defend your answer.

Figure 23-8: How are paleontologists able to reconstruct an entire skeleton (and, in turn, an entire "body") from fragments such as those shown in this figure?

Figure 23-9: Why don't the shadows of ancestral species overlap with the descendants?

Why are the australopithecines represented across the diagram in the two through four million year range?

Why is *Homo erectus* at the same end of the diagram as *Homo habilis* and not nearer the australopithecines?

Figure 23-10: How can paleontologists determine the age of these footprints?

## Matching Questions

Write the letter of the phrase that best matches the numbered term on the left. Use each only once.

____ 1. hominid

____ 2. homologous feature

____ 3. analogous feature

____ 4. primitive feature

____ 5. derived feature

____ 6. fossil record

____ 7. comparative embryology

____ 8. biogeography

____ 9. vestigial structure

____10. *Archaeopterix*

____11. australopithecines

____12. Piltdown Man

____13. *Homo habilis*

____14. *Australopithecus afarensis*

____15. *Homo erectus*

____16. *Homo sapiens*

____17. Olduvai Gorge

____18. Eugene Dubois

____19. Raymond Dart

____20. Louis and Mary Leakey

____21. Donald Johanson

a. similarity of two species because of convergent evolution

b. a reduced organ that has no apparent function in an organism

c. the study of distribution of organisms

d. a group of early hominids considered by many not to be on the lineage leading to modern humans

e. a hominid species that lived 3.5 million years ago

f. responsible for naming *A. africanus*

g. study of developmental similarities and differences between species

h. scientific name for modern humans

i. discovered "Lucy"

j. remains used by paleontologists to piece together evolutionary history

k. similarity of two species because of a common ancestor

l. discovered a molar from "Java Man"

m. a fossil forgery

n. a fossil bird with many reptilian characteristics

o. species of Peking Man and Java Man

p. the earliest known hominid tool user

q. anthropological site in East Africa

r. modified from the ancestral arrangement

s. general term applied to humans and their recent ancestors that walked upright

t. discovered *Homo habilis*

u. the arrangement seen in the ancestor

## Multiple Choice Questions

1. Structures which have a similar construction and embryonic origin, but not necessarily a similar function are said to be
   a. homoplastic.
   b. analogous.
   c. deleterious.
   d. homologous.
   e. analytical.

2. Structures that look alike due to a similar function but not because of common ancestry are said to be
   a. analogous.
   b. homologous.
   c. homosporous.
   d. analgesic.
   e. anagogic.

3. The current evolutionary model suggests that fish, frogs, cats and humans had a common ancestor that sets them apart from other animals. Which primitive characteristic is consistent with this?
   a. they are composed of cells
   b. they perform glycolysis
   c. they perform the Krebs cycle
   d. they are eukaryotes
   e. they have a backbone

4. The current evolutionary model suggests that sea lions and sharks independently evolved as marine organisms. Which derived characteristic supports this?
   a. sharks have paddle-shaped appendages, sea lions do not
   b. sea lions have paddle-shaped appendages, sharks do not
   c. sharks have a backbone, sea lions do not
   d. sea lions have hair, sharks do not
   e. sharks are made of cells, sea lions are not

5. Lions and tigers belong to the same mammalian order (Carnivora) and animal subphylum (Vertebrata). Which is most likely true of lions and tigers?
   a. The cat-like similarities are due to convergent evolution.
   b. The differences are due to divergent evolution.
   c. They are more closely related to elephants (order Proboscidea) than to each other.
   d. Their vertebral columns are analogous structures.
   e. They share a common ancestor with stripes and a mane.

6. Whales and dolphins (small whales) belong to the same order of aquatic mammals (Cetacea). Which is most likely true of whales and dolphins?
   a. The similarities of a fish-like body are due to convergent evolution.
   b. The differences are due to divergent evolution.
   c. They are more closely related to seals (order Pinnipedia) than to each other.
   d. Their lungs are analogous structures.
   e. They do not share a common ancestor with fish-like body features.

7. Many members of the plant family Euphorbiaceae are cactus-like in appearance. True cacti belong to the family Cactaceae. Both are angiosperm families. Which most likely describes euphorbs and cacti?
   a. The spines are the result of convergent evolution.
   b. The presence of flowers in both is the result of convergent evolution.
   c. The presence of xylem and phloem is the result of convergent evolution.
   d. The stamens in both are analogous structures.
   e. The roots are analogous structures.

8. Penguins are birds (class Aves) and seals are mammals (class Mammalia). Both belong to the phylum Chordata and have a vertebral column surrounding a spinal cord. Which is true of penguins and seals?
   a. The fins are the product of convergent evolution.
   b. The presence of the vertebral column is the result of convergent evolution.
   c. The flippers are homologous structures.
   d. The streamlined body shapes are homologous structures.
   e. The spinal cords are analogous structures.

9. Sharks are fish (class Chondryichthyes) and have gills, whereas dolphins are mammals (class Mammalia) and have lungs. Both belong to the phylum Chordata and have a vertebral column. Which is most likely true of sharks and dolphins?
   a. The vertebral columns are analogous structures.
   b. The dorsal fins are homologous structures.
   c. The tail fins are the product of convergent evolution.
   d. Lungs and gills are analogous structures.
   e. Lungs and gills are homologous structures.

10. Whales belong to the mammalian order Cetacea, while seals belong to the mammalian order Pinnipedia. The evidence indicates that both evolved from different terrestrial mammalian ancestors. Which is most likely true of whales and seals?
    a. The front fins are homologous structures.
    b. The forelimb bones are homologous structures.
    c. The presence of hair follicles is the result of convergent evolution.
    d. The ability to maintain a constant body temperature is the result of convergent evolution.
    e. The ability to breathe air is the result of convergent evolution.

11. Compared to other mammals, which describes seals (class Mammalia, order Pinnipedia)?
    a. Fur is a derived characteristic.
    b. The presence of lungs is a derived characteristic.
    c. The vertebral column is a primitive characteristic.
    d. The presence of flippers is a primitive characteristic.
    e. The fish-like tail is a primitive characteristic.

12. Compared to other flowering plants, which describes cacti (division Anthophyta, family Cactaceae)?
    a. Flowers are derived structures.
    b. Xylem and phloem are primitive structures.
    c. Roots are derived structures.
    d. Chloroplasts are derived structures.
    e. Spines (modified leaves) are primitive structures.

13. Fossils are often found in layers of sediment ("strata"). Where should the oldest fossils be found in undisturbed strata?
    a. in the uppermost stratum
    b. in the lowest stratum
    c. near the top, but not on the top
    d. near the bottom, but not on the bottom
    e. uniformly distributed throughout the strata

14. Radioisotopes provide a means of
    a. determining absolute age of fossils.
    b. determining relative age of fossils.
    c. distinguishing between homologous structures.
    d. distinguishing between analogous structures.
    e. distinguishing between birds and reptiles.

15. Which is most likely to become fossilized?
    a. herbaceous plant
    b. flower
    c. tooth
    d. jelly fish
    e. earthworm

16. Which is most likely to become fossilized?
    a. a desert insect
    b. a marine clam
    c. a flowering plant from the mountains
    d. prairie grass
    e. a desert snake

17. According to current evidence, which is the oldest species?
    a. *Homo habilis*
    b. *Homo erectus*
    c. *Homo sapiens*
    d. *Australopithecus robustus*
    e. *Australopithecus afarensis*

18. According to current evidence, which is the youngest species?
    a. *Homo habilis*
    b. *Homo erectus*
    c. *Homo sapiens*
    d. *Australopithecus robustus*
    e. *Australopithecus afarensis*

19. According to current evidence, in which sequence did these hominids first appear?
    a. *Homo habilis, Homo erectus, Homo sapiens, Australopithecus africanus, Australopithecus afarensis*
    b. *Homo erectus, Homo habilis, Homo sapiens, Australopithecus africanus, Australopithecus afarensis*
    c. *Homo sapiens, Homo habilis, Homo erectus, Australopithecus africanus, Australopithecus afarensis*
    d. *Australopithecus africanus, Australopithecus afarensis, Homo habilis, Homo erectus, Homo sapiens*
    e. *Australopithecus afarensis, Australopithecus africanus, Homo habilis, Homo erectus, Homo sapiens*

20. In which sequence were these hominids discovered?
    a. *Homo habilis, Homo erectus, Australopithecus africanus, Australopithecus afarensis*
    b. *Homo erectus, Australopithecus africanus, Homo habilis, Australopithecus afarensis*
    c. *Homo habilis, Australopithecus africanus, Australopithecus afarensis, Homo erectus*
    d. *Homo erectus, Australopithecus africanus, Australopithecus afarensis, Homo habilis,*
    e. *Australopithecus afarensis, Australopithecus africanus, Homo habilis, Homo erectus*

## Concept Map Construction

Construct a concept map for each group of terms. Be sure to include appropriate connector phrases. You may add other terms as necessary and use terms in the singular or plural form.

1. homology, convergent evolution, vestigial structure, comparative embryology, biogeography
2. *Homo habilis*, gene pool, natural selection, fossil record, adaptive radiation
3. reproductive isolation, bone, radioactive carbon, species, *Australopithecus afarensis*

# Chapter 24

# Evolution: Origin and History of Life

## Go Figure!

Figure 24-1: How might biologists recognize the organism in Figure 24-1a as a filamentous cyanobacterium?

How do biologists know the filamentous cyanobacterium in Figure 24-1a did not die recently?

How can biologists assign an age of 3.5 billion years to a specimen?

Figure 24-2: How much longer was the Proterozoic Era than the other three eras combined?

Why do geologists cram three eras in a time span that is only a fraction as long as the Proterozoic?

What characteristics of Proterozoic organisms might make it difficult for geologists to break this era into smaller time spans?

What correlation between existing species diversity, competition intensity, and evolutionary rate is suggested by this chart?

Figure 24-3: Why are fossils of soft-bodied organisms relatively rare?

The age of this fossil is given as approximately 650 million years. How are biologists able to give estimates of fossil age in years?

How many years represents a 1% error in the estimate of this fossil's age?

Figure 24-4: For at least how many years has North America been separated from Africa?

At what point did India separate from Gondwanaland?

Was the joining of India and Asia a recent geologic event?

Figure 24-5: Which vertebrate groups (fish, amphibians, reptiles, birds, or mammals) first appeared during the Paleozoic Era?

Which major plant groups (nonseed plants, gymnosperms, and flowering plants) first appeared during the Paleozoic era?

Assume the first vascular plants appeared at the beginning of the Silurian Period. What percent of earth's existence have vascular plants been living?

Figure 24-6: Assume mammals first appeared at the beginning of the Triassic Period. What percent of earth's existence have mammals been living?

Dinosaurs were dominant land animals for 125 million years. What percent of earth's existence does this represent?

Figure 24-7: Why were teeth, jaws and upright posture used in identifying the therapsid's mammalian relationship?

Why weren't such important mammalian features such as mammary glands, fur, and constant body temperature not used?

Figure 24-8: In Chapter 23 we saw that *Australopithecus afarensis* ("Lucy") lived 3.5 million years ago. During which geologic period was this?

Assume modern humans (*Homo sapiens*) first appeared on earth 200,000 years ago (currently the earliest estimate of our arrival). Life first appeared 3.5 billion years ago. What percentage of that time have modern humans been living?

## Matching Questions

Write the letter of the phrase that best matches the numbered term on the left. Use each only once.

____ 1. chemical evolution

____ 2. 3.5 billion years

____ 3. prokaryotic cells

____ 4. mass extinction

____ 5. cyanobacteria

____ 6. Oparin

____ 7. Pangaea

____ 8. Laurasia

____ 9. Gondwanaland

____10. anaerobic photosynthetic bacteria

____11. Mesozoic Era

____12. Proterozoic Era

____13. Cenozoic Era

____14. Pleistocene Epoch

____15. Paleozoic

____16. therapsid

____17. stromatolite

____18. pterosaur

a. product of Pangaea containing Africa

b. reptile group thought to be ancestral to the mammals

c. group responsible for producing the aerobic atmosphere

d. characterized by periods of glaciation

e. first flying vertebrate

f. reptiles dominated this time in history

g. a mass of prokaryotic cells and minerals existing today, but also found as fossil evidence of early life

h. mammals dominate this time in history

i. the formation of life from the development of increasingly complex molecules

j. these gave rise to eukaryotic cells 2 billion years ago

k. Russian biochemist who proposed the idea of chemical evolution

l. often precedes major periods of adaptive radiation

m. geologic time period in which the first cells evolved

n. first autotrophs

o. a supercontinent that existed 200 million years ago

p. geologic time period in which first seed plants evolved

q. product of Pangaea containing North America

r. age of the oldest fossils

## Multiple Choice Questions

1. Which DOES NOT describe the formation of the sun?
   a. It formed from matter.
   b. The gravitational attraction between particles caused it to form.
   c. Thermonuclear reactions occurred as a result of particles coming together.
   d. The sun formed 3.5 million years ago.
   e. The sun formed at the center of the solar system.

2. Which DOES NOT describe the formation of the earth?
   a. Its atmosphere vaporized due to intense heat from the sun.
   b. It condensed from the same matter that produced the sun.
   c. The earth was initially very hot.
   d. It formed at the same time as the sun.
   e. Heavier elements sank to its core.

3. Which DOES NOT describe the Proterozoic Era?
   a. The first multicellular organisms were formed.
   b. The first cells arose.
   c. Oxygen accumulated in the atmosphere.
   d. It is the shortest of the four geologic eras.
   e. The first prokaryotes arose.

4. Which DOES NOT describe the Proterozoic Era?
   a. Competition for resources and natural selection did not occur.
   b. The atmosphere initially lacked oxygen.
   c. Chemical evolution occurred.
   d. It covers a four billion year time period beginning with the origin of the earth.
   e. The first eukaryotes arose.

5. The first heterotrophs lived during the
   a. Proterozoic Era.
   b. Paleozoic Era.
   c. Mesozoic Era.
   d. Cambrian Period.
   e. Jurassic Period.

6. The first autotrophs lived during the
   a. Pleistocene Epoch.
   b. Triassic Period.
   c. Proterozoic Era.
   d. Devonian Period.
   e. Permian Period.

7. Of these five, which came first?
   a. reptiles
   b. mammals
   c. autotrophic prokaryotes
   d. multicellular eukaryotes
   e. heterotrophic prokaryotes

8. Of these five, which came first?
   a. anaerobic respiration
   b. aerobic respiration
   c. anaerobic photosynthesis
   d. aerobic photosynthesis
   e. fermentation

9. Which DOES NOT describe multicellularity?
   a. Evidence suggests that it developed only once.
   b. It is found only in eukaryotic organisms.
   c. It enables cells to perform specialized functions.
   d. Some multicellular organisms arose from colonies of cells.
   e. Multicellular organisms first appeared in the Proterozoic Era.

10. Which DOES NOT describe multicellularity?
    a. Some multicellular organisms arose from aggregations of independent cells.
    b. Multicellular organisms arose about one billion years ago.
    c. Multicellular prokaryotes gave rise to multicellular eukaryotes in the late Proterozoic Era.
    d. Multicellularity preceded the development of the major animal phyla.
    e. Multicellularity preceded the development of the major plant phyla.

11. Which describes the theory of continental drift?
    a. It has been replaced by more modern models of chemical evolution.
    b. It may explain worldwide climate changes.
    c. It explains the origin of life.
    d. It can account for unusual fossil distributions.
    e. It states that Pangaea split into Laurasia and Magma about 150 million years ago.

12. Which is consistent with the theory of continental drift?
    a. Mountains may be built when continental plates collide.
    b. Volcanoes cause continental plates to move.
    c. Though they have moved, continents have always been in their current relative positions with respect to each other.
    d. Continents have moved, but they have always been separate.
    e. There is no evidence of continental movement, just expansion of the oceans.

13. Of these five, which came first?
    a. reptiles
    b. fish
    c. amphibians
    d. birds
    e. mammals

14. Of these five, which came first?
    a. algae
    b. land plants
    c. vascular plants
    d. seed plants
    e. flowering plants

15. The first land plants lived during the
    a. Paleozoic Era
    b. Jurassic Period
    c. Cambrian Period
    d. Silurian Period
    e. Triassic Period

16. The first vertebrates lived during the
    a. Mesozoic Era
    b. Permian Period
    c. Ordovician Period
    d. Devonian Period
    e. Pleistocene Epoch

17. Dinosaurs were the dominant land animals during the
    a. Jurassic Period
    b. Triassic Period
    c. Cretaceous Period
    d. Tertiary Period
    e. Quaternary Period

18. Mammals became the dominant land mammals during the
    a. Jurassic Period
    b. Triassic Period
    c. Cretaceous Period
    d. Tertiary Period
    e. Quaternary Period

19. Gymnosperms were the dominant land plants during the
    a. Cambrian Period
    b. Triassic Period
    c. Cretaceous Period
    d. Ordovician Period
    e. Permian Period

20. Flowering plants first appeared during the
    a. Cambrian Period
    b. Triassic Period
    c. Cretaceous Period
    d. Ordovician Period
    e. Permian Period

## Concept Map Construction

Construct a concept map for each group of terms. Be sure to include appropriate connector phrases. You may add other terms as necessary and use terms in the singular or plural form.

1. divergent evolution, chemical evolution, geologic time scale, origin of life, prokaryote
2. *Homo sapiens*, continental drift, fossil, therapsid, adaptive radiation
3. autotroph, homeostasis, dinosaur, mass extinction, natural selection

# Chapter 25

# The Microbial Kingdoms: Viruses, Monerans, Protists, and Fungi

## Go Figure!

Figure 25-1: Which structures are present in both viruses and bacteria?

Which bacterial structures are absent from viruses?

What is the molecular basis for specificity between virus particles and their hosts?

Figure 25-2: Which part of the virus enters the host cell?

Figure 25-3: The bacteria in Figure 25-3a form clusters. Why are these bacteria not considered by biologists to be multicellular organisms?

Figure 25-4: Why do the authors suggest that these modern anaerobic species of bacteria might be similar to the earliest forms of life?

What does the fact that *Clostridium* species live in places such as canned food and wounded flesh suggest about their type of metabolism?

Figure 25-5: The caption title for this figure is "One proposal of evolutionary descent of all organisms from early prokaryotic cells." What does that tell you about the current state of classification at the kingdom level?

Is it possible to have different descent patterns where each is consistent with available evidence?

Which groups belong to the eubacteria?

What evidence supports the claim that archaebacteria are not descendents of other prokaryotes?

Figure 25-6: What cell shapes are shown in these photomicrographs?

What are the main functions of bacterial stains?

Figure 25-7: Why should the results of biochemical tests run on a single colony of bacteria be identical to the results run on any single cell in the same colony?

Figure 25-8: Which structures shown do you consider essential for a bacterium to be minimally functional? Why?

Figure 25-9: Prokaryotes lack membranous organelles. Why are the stacks of internal membranes in this cyanobacterium not considered to be membranous organelles?

What membrane are they made of?

Figure 25-10: The nitrogen-fixing enzyme is sensitive to oxygen and will not function in its presence. Which part(s) of the photosynthetic apparatus in a heterocyst must be disabled to prevent oxygen inactivation of the enzyme?

Figure 25-13: Why is *Euglena* placed in the protist kingdom rather than the plant kingdom?

What characteristics are visible in the picture that eliminate *Euglena* from the kingdom Monera?

Figure 25-14: If these two *Euplotes* simply exchange genetic material, is this an example of sexual reproduction?

Figure 25-15: Your text suggests that multicellular algae are placed in the protist kingdom by some authorities and in the plant kingdom by others. How is this issue addressed in the illustration?

Figure 25-16: To which of the four protozoan groups shown in Figure 25-15 do these organisms belong?

Figure 25-17: How many hosts are used during the *Plasmodium* life cycle?

Why are public health agencies concerned with mosquito control?

Figure 25-19: What characteristic shown in the picture indicates the organism is a dinoflagellate?

Figure 25-25: How does dikaryotic differ from diploid?

What process produces basidiospores?

Is kicking over a mushroom in your lawn an effective way of ridding your lawn of the fungus?

## Matching Questions

Write the letter of the phrase that best matches the numbered term on the left. Use each only once.

____ 1. eubacteria
____ 2. nucleoid
____ 3. peptidoglycan
____ 4. binary fission
____ 5. cyanobacteria
____ 6. heterocyst
____ 7. methanogens
____ 8. saprophyte
____ 9. virus
____10. pseudopodia
____11. vaccine
____12. zooplankton
____13. trophozoite
____14. cyst
____15. diatoms
____16. dinoflagellates
____17. cellular slime mold
____18. plasmodial slime mold
____19. zygospore
____20. basidiospore
____21. deuteromycete

a. a structure used for nitrogen fixation
b. life cycle includes an amoeba-like stage and a fruiting stage
c. one part of its life cycle includes a large, multinucleated "cell"
d. cell wall material of most prokaryotes
e. these are characterized by having two flagella
f. the *Plasmodium* stage found in red blood cells of the host
g. the group of true bacteria
h. used in amoeboid motility
i. fungus in which no evidence of sexual reproduction has been discovered
j. these prokaryotes produce oxygen during photosynthesis
k. a general term for a noncellular parasite
l. type of asexual reproduction
m. DNA region of a prokaryotic cell
n. archaebacteria that produce methane gas
o. an organism that gets nutrients from dead matter
p. stimulates the immune system prior to infection by a pathogenic virus
q. the dormant form of a polymorphic protozoan
r. algae with silica cell walls
s. produced in mushrooms
t. only diploid cell in the zygomycete life cycle
u. a group of unicellular heterotrophs

## Multiple Choice Questions

1. Which IS NOT a characteristic of the Kingdom Monera?
   a. peptidoglycan cell wall
   b. prokaryotic ribosomes
   c. rapid growth rate
   d. a variety of metabolic capabilities
   e. true nucleus

2. Which IS NOT a characteristic of the Kingdom Monera?
   a. small size
   b. large population size
   c. asexual reproduction by fission
   d. presence of endoplasmic reticulum
   e. a variety of cell shapes

3. *Escherichia coli* can divide every 20 minutes if grown in optimum conditions. If you start with one cell in optimum conditions at 8 AM and it divides after 20 minutes, how many cells will you have at 1:30 PM that same day? (Assume no cells die and each cell divides on a synchronous 20 minute schedule.)
   a. 17
   b. 34
   c. 512
   d. 65,536
   e. 131,072

4. Suppose an *E. coli* cell is grown in an environment that changes the time between divisions to 30 minutes. If you start with one cell at 8 AM, how many cells will you have at 1:20 PM that same day? (Assume no cells die and each cell divides on a synchronous 30 minute schedule.)
   a. 2048
   b. 1024
   c. 512
   d. 128
   e. 11

5. Which DOES NOT describe the Eubacteria?
   a. It includes the majority of prokaryotic species.
   b. Eubacteria demonstrate great metabolic diversity.
   c. Photosynthetic eubacteria may use bacterial chlorophyll.
   d. Eubacteria have a protein cell wall.
   e. Some pathogenic eubacteria form a capsule that impairs host defenses.

6. Which DOES NOT describe the Eubacteria?
   a. Some produce endospores.
   b. Most eubacteria are pathogenic.
   c. Flagella act as rotors to produce movement.
   d. Eubacteria lack a membrane-bound nucleus.
   e. Cyanobacteria are considered to be eubacteria.

7. Which describes cyanobacteria?
   a. They are halophilic.
   b. All cells are capable of nitrogen fixation.
   c. They lack chloroplasts.
   d. They have a cellulose cell wall.
   e. They are strictly unicellular.

8. Which describes cyanobacteria?
   a. They are classified with the other algae.
   b. Heterocysts are cells specialized for photosynthesis.
   c. They are primarily pathogenic.
   d. They always form filaments of cells.
   e. Their photosynthesis produces oxygen.

9. Which describes the Archaebacteria?
   a. Their cell walls are made of peptidoglycan.
   b. They often inhabit extreme environments.
   c. They have heterocysts that allow them to fix $CO_2$.
   d. They are eukaryotes.
   e. They include the majority of Monerans.

10. Which describes the Archaebacteria?
    a. They are urkaryotes.
    b. They include cyanobacteria.
    c. Some are thermophilic.
    d. They have typical membranes made of phospholipids.
    e. They have a nuclear membrane around DNA.

11. Which DOES NOT describe viruses?
    a. Each virus has both DNA and RNA in it.
    b. Viruses are obligate intracellular parasites.
    c. They exhibit some, but not all, of the characteristics associated with living organisms.
    d. Replication of a single virus may result in production of hundreds of new viruses.
    e. A protein coat surrounds the nucleic acid portion of each virus.

12. Which DOES NOT describe viruses?
    a. They are larger than archaebacteria.
    b. The replicative cycle includes a period in which no complete virus particles are present.
    c. They infect host cells.
    d. Host cell metabolic machinery is used during viral replication.
    e. They are not cellular organisms.

13. What is the most effective protection against viral infections?
    a. disinfectants
    b. natural immunity of the host
    c. antibiotics
    d. analgesics
    e. washing hands

14. Why are most chemical treatments generally ineffective at combating viral infections?
    a. Viruses are not made of chemicals, so chemical treatments are ineffective.
    b. Viruses are not made of the same chemicals as living organisms, so chemical treatments are ineffective.
    c. Viruses enter host cells and it is difficult to attack them without damaging the host.
    d. Viruses have enzymes that break down most antiviral chemicals.
    e. Viruses have a selectively permeable membrane that prevents entry of most antiviral chemicals.

15. Which is an evolutionary advancement credited to protists?
    a. fermentation
    b. photosynthesis
    c. DNA replication
    d. sexual reproduction
    e. ribosomes

16. Which is an evolutionary advancement credited to protists?
    a. photosynthesis
    b. chloroplasts
    c. cell membrane
    d. cell division
    e. motility

17. Which describes a "typical" protozoan?
    a. multicellular
    b. cellulose cell wall
    c. photosynthetic
    d. ingestive heterotroph
    e. nonmotile

18. Which describes a "typical" protozoan?
    a. peptidoglycan cell wall
    b. chitinous cell wall
    c. autotrophic
    d. pathogenic to humans
    e. motile

19. To which group does a unicellular, photosynthetic protist with two flagella belong?
    a. blue-green algae
    b. dinoflagellate
    c. diatom
    d. ciliate
    e. Sarcodina

20. To which group does a single celled, photosynthetic protist with silica cell walls belong?
    a. blue-green algae
    b. dinoflagellate
    c. diatom
    d. flagellate
    e. ascomycete

21. Which DOES NOT describe cellular slime molds?
    a. One stage of the life cycle involves amoeboid cells.
    b. They are heterotrophic.
    c. Individual cells come together to form a fruiting body when environmental conditions are poor.
    d. Cyclic AMP is a chemical used to organize development during one stage of the life cycle.
    e. Spore germination produces a fruiting body.

22. Which DOES NOT describe plasmodial slime molds?
    a. One stage of the life cycle involves amoeboid cells.
    b. They are heterotrophic.
    c. One stage of the life cycle involves a large, multinucleate cell.
    d. The plasmodium is the feeding stage of the life cycle.
    e. The plasmodium develops from a zygote.

23. Which describes fungi?
    a. Most are autotrophic.
    b. Most are motile.
    c. Most are unicellular.
    d. Most are parasitic.
    e. Most are capable of sexual reproduction.

24. Which describes fungi?
    a. Typically, the only diploid cell is the zygote.
    b. Photosynthetic fungi are important members of terrestrial food chains.
    c. Unicellular fungi form hyphae when mature.
    d. Haploid spores combine to form a diploid zygote.
    e. Spores are the only haploid cells in most fungal life cycles.

25. Which terrestrial fungal group is characterized by nonseptate hyphae and the absence of flagella?
    a. ascomycetes
    b. zygomycetes
    c. basidiomycetes
    d. deuteromycetes
    e. oomycetes

26. Which terrestrial fungal group is characterized by sexual spores contained in an ascus?
    a. ascomycetes
    b. zygomycetes
    c. basidiomycetes
    d. deuteromycetes
    e. oomycetes

27. Which terrestrial fungal group is responsible for producing mushrooms and puffballs?
    a. ascomycetes
    b. zygomycetes
    c. basidiomycetes
    d. deuteromycetes
    e. oomycetes

28. Which terrestrial fungal group has no apparent sexual component in the life cycle?
    a. ascomycetes
    b. zygomycetes
    c. basidiomycetes
    d. deuteromycetes
    e. oomycetes

29. Which describes basidiospores?
    a. They are diploid and produced by meiosis.
    b. They are haploid and produced by meiosis.
    c. They are diploid and produced by mitosis.
    d. They are haploid and produced by mitosis.
    e. They are dikaryotic and produced by fertilization.

30. Which describes ascospores?
    a. They are diploid and produced by meiosis.
    b. They are haploid and produced by meiosis.
    c. They are diploid and produced by mitosis.
    d. They are haploid and produced by mitosis.
    e. They are dikaryotic and produced by fertilization.

## Concept Map Construction

Construct a concept map for each group of terms. Be sure to include appropriate connector phrases. You may add other terms as necessary and use terms in the singular or plural form.

1. binary fission, metabolism, chemosynthesis, cyanobacteria, heterocyst
2. prokaryote, cell wall, photosynthesis, phytoplankton, dinoflagellate
3. heterotroph, lichen, saprobe, basidiospore, deuteromycete

# Chapter 26

# The Plant Kingdom

## Go Figure!

Figure 26-1: Where do plants alive today appear on the diagram?

Why might this be drawn with two lines leading to the bryophytes?

Figure 26-2: What do the colored circles represent?

What do the different amounts of colored space in each circle represent?

In this context, what does "prominent" mean?

What process allows the 2N sporophyte to become haploid (1N)?

What process allows the haploid generation to become diploid again?

Figure 26-3: Which diagram shows bryophytes and tracheophytes with green algae as a common ancestor?

Why isn't the hypothesis of bryophytes giving rise to tracheophytes considered as a third option?

Figure 26-4: To which division does each organism belong?

Figure 26-5: With the exception of the objects that look like angry cobras in the moss photograph (Figure 26-5a), all the plants are haploid. Which generation is shown?

What moss generation do the "angry cobras" represent?

Are they haploid or diploid?

Figure 26-6: What structure surrounds the zygote?

What process transforms the zygote into a multicellular sporophyte?

How does the single celled spore develop into a multicellular gametophyte?

Figure 26-7: How can you tell this is a photograph of *Psilotum* sporophytes?

Figure 26-9: How can you tell this is a photograph of a *Lycopodium* sporophyte?

What plant organ does *Lycopodium* have that *Psilotum* doesn't?

Figure 26-11: Observe the part labelled "Developing sporophyte." The vertical plant part is the first sporophyte leaf; the sporophyte root is projecting downwards. What is the horizontal part?

Figure 26-12: Why are cycads considered gymnosperms when many resemble palms?

Figure 26-15: What process produces cones?

What process produces microspores?

What process produces megaspores?

Which gamete is contained inside pollen grains?

At which point on the diagram is the zygote found?

What process produces an embryo?

What process produces a mature sporophyte?

Figure 26-16: What dicot characteristics are visible in pictures Figures 26-16a and 26-16b?

What monocot characteristics are visible in pictures Figures 26-16c and 26-16d? ("None" is an acceptable answer!)

## Matching Questions

Write the letter of the phrase that best matches the numbered term on the left. Use each only once.

____ 1. sporophyte generation

____ 2. gametophyte generation

____ 3. cuticle

____ 4. stomate

____ 5. chlorophyta

____ 6. bryophyta

____ 7. fucoxanthin

____ 8. antheridium

____ 9. archegonium

____10. strobilus

____11. flagellated sperm

____12. frond

____13. prothallus

____14. coniferophyta

____15. sphenophyta

____16. pterophyta

____17. epiphyte

____18. microspore

____19. seed

____20. fruit

a. retards water loss by a terrestrial plant

b. pigment characteristic of phaeophytes

c. develops into a pollen grain

d. bryophyte sperm producing structure

e. fern leaf

f. horsetails belong to this division

g. develops from an ovule, contains an embryo and is characteristic of higher vascular plants

h. pores in the surface of a terrestrial plant's body

i. develops from an angiosperm ovary

j. most obvious generation of higher vascular plants

k. fern gametophyte

l. most common gymnosperm division

m. plant group composed of mosses and liverworts

n. a plant that grows on another plant, but is not parasitic

o. bryophyte egg producing structure

p. the division of ferns

q. the division of green algae

r. a cluster of sporophylls in certain vascular plants

s. most obvious generation of lower vascular plants

t. gamete typical of ferns

## Multiple Choice Questions

1. Which is an adaptation of plants to a terrestrial environment?
   a. photosynthesis
   b. fermentation
   c. cuticle
   d. chlorophyll
   e. gametes

2. Which is an adaptation of plants to a terrestrial environment?
   a. xylem and phloem
   b. cellular respiration
   c. chloroplasts
   d. motile sperm
   e. mitochondria

3. Which describes the chlorophyta?
   a. This division includes all algae.
   b. It is thought that an ancient chlorophyte is ancestral to true plants.
   c. Fucoxanthin is the primary photosynthetic pigment.
   d. Most are marine plants.
   e. Members have a peptidoglycan cell wall.

4. Which describes the chlorophyta?
   a. There is no functional difference between sporophytes and gametophytes.
   b. Male and female gametes are indistinguishable in all species.
   c. The pigments chlorophyll *a* and *b* are present in all species.
   d. Glycogen is the storage carbohydrate.
   e. Members are unicellular or colonial, but never multicellular.

5. Which describes the phaeophyta?
   a. Growth rates of all species are very slow.
   b. The plants tend to grow in shallow water.
   c. There is no cellular differentiation in any species.
   d. Species have either a sporophyte generation or a gametophyte generation, never both.
   e. This is the division of brown algae.

6. Which describes the phaeophyta?
   a. The group is important ecologically, but has no commercial importance.
   b. Large species may have holdfasts, blades and floats.
   c. The gametophyte is the obvious generation of large species.
   d. The most important accessory pigment is rhodopsin.
   e. Some plants are the source of agar.

7. Which describes the rhodophyta?
   a. It is thought that an ancient rhodophyte is ancestral to true plants.
   b. The members are predominantly found in freshwater environments.
   c. They are important members of the large kelp forests.
   d. Because of their pigments, they are able to survive in deeper waters.
   e. They are commonly called brown algae.

8. Which describes the rhodophyta?
   a. This is the division of green algae.
   b. The members may reproduce sexually or asexually.
   c. No sexual stage in the life cycle has been discovered as yet.
   d. No sporophyte stage in the life cycle has been discovered as yet.
   e. Most members live in terrestrial habitats.

9. Which describes the bryophyta?
   a. Bryophytes have no cellular specialization.
   b. Bryophytes one meter or more in height are common.
   c. Bryophytes exhibit alternation of generations.
   d. Bryophyte seeds are capable of remaining dormant for long periods of time.
   e. Bryophytes are not capable of sexual reproduction.

10. Which describes the bryophyta?
    a. The diploid spore producing generation is most obvious.
    b. Water is necessary for bryophyte sperm to swim in on their journey to the egg.
    c. The gametophyte is dependent on the sporophyte for nutrition.
    d. Individuals are always bisexual.
    e. Taller bryophytes have xylem and phloem.

11. During a morning stroll on the beach, you discover a large brownish plant with leafy structures washed up onto shore. You take it back to your laboratory and discover it has fucoxanthin pigment in its cells. To which division does this plant belong?
    a. bryophyta
    b. chlorophyta
    c. phaeophyta
    d. rhodophyta
    e. xanthophyta

12. While exploring in a forest, you find a small greenish plant growing on a rock near a stream. Microscopic examination reveals haploid cells and multicellular gametangia. Chemical analysis indicates starch and chlorophyll *a* and *b* are present. To which division does this plant belong?
    a. bryophyta
    b. chlorophyta
    c. phaeophyta
    d. rhodophyta
    e. xanthophyta

13. Which describes the pterophyta?
    a. The leaves are small and inconspicuous.
    b. Sporangia are grouped into sori.
    c. They do not require water for fertilization.
    d. Sporangia are found on the prothallus.
    e. The gametophyte generation is most obvious.

14. Which describes the pterophyta?
    a. They were the first vascular plants to produce seeds.
    b. Gametophytes have vascular tissue.
    c. Insects play an important role in transfer of spores between plants.
    d. The most obvious stage of the life cycle is diploid.
    e. This is the division of whisk ferns.

15. Which is a gymnosperm advancement over lower vascular plants?
    a. Gymnosperms protect sperm cells in a dry pollen grain.
    b. Gymnosperms protect the embryo in a fruit.
    c. Gymnosperms produce xylem and phloem.
    d. Gymnosperms produce roots, stems and leaves.
    e. Gymnosperms produce spores.

16. Which is a gymnosperm advancement over lower vascular plants?
    a. Gymnosperms produce gametes.
    b. Gymnosperms produce motile sperm.
    c. Gymnosperms are able to perform cellular respiration.
    d. Gymnosperms are able to photosynthesize.
    e. Gymnosperms protect the embryo in a seed.

17. Which is a gymnosperm division?
    a. psilophyta
    b. cycadophyta
    c. anthophyta
    d. chrysophyta
    e. pyrrophyta

18. Which is a gymnosperm division?
    a. ginkgophyta
    b. sphenophyta
    c. chlorophyta
    d. cyanophyta
    e. tracheophyta

19. Which describes pines?
    a. They belong to the smallest gymnosperm division.
    b. Most species have male and female cones on different trees.
    c. All species have woody stems.
    d. Pollen is produced in female cones.
    e. It takes a single growing season to go from production of a female cone to seed formation.

20. Which describes pines?
    a. Each cone contains antheridia for sperm production.
    b. Sperm have two flagella for motility.
    c. Megaspore mother cells divide to produce male gametophytes.
    d. Pollen grains have structures that aid in wind dispersal.
    e. Cones are produced by conifer gametophytes.

21. Which is an angiosperm advancement over gymnosperms?
    a. Angiosperms do not require water for fertilization.
    b. Angiosperms produce flowers to attract specific pollinators.
    c. Angiosperms produce pollen.
    d. Angiosperms have a dominant gametophyte generation.
    e. Angiosperms have woody stems.

22. Which is an angiosperm advancement over gymnosperms?
    a. Angiosperms require water for fertilization.
    b. Angiosperms produce seeds.
    c. Angiosperms produce cones.
    d. Angiosperms have a dominant sporophyte generation.
    e. Angiosperms produce fruit for seed dispersal.

## Concept Map Construction

Construct a concept map for each group of terms. Be sure to include appropriate connector phrases. You may add other terms as necessary and use terms in the singular or plural form.

1. alternation of generations, fertilization, chlorophyta, sperm, spore
2. psilophyta, pterophyta, xylem, sorus, meiosis
3. coniferophyta, monocot, flower, cone, sporophyte

# Chapter 27

# Animal Diversity

## Go Figure!

Figure 27-1: Why are such obviously different organisms as *Trichoplax adhaerans* and a blue whale considered members of the same kingdom?

At what taxonomic level are the two separated?

Figure 27-4: What is the function of porocytes?

What is the function of choanocytes?

How is increased surface area for choanocytes an advantage?

Figure 27-5: Which organisms are polyps?

Are any in the medusa form?

What is the fundamental difference between polyps and medusas?

What kind of symmetry does each have?

Figure 27-6: Why are cnidarians classified as only having a tissue level of organization?

What are those tissues?

Where are cnidocytes concentrated in each body form?

What is the role of cnidocytes?

Figure 27-7: What kind of symmetry does the flatworm have?

Figure 27-9: How many different organ systems are represented in this "simple" worm?

How does a pseudocoelom differ from a true coelom?

Figure 27-10: How many different organ systems are represented in this drawing of a snail?

Which characteristics are distinctly molluscan?

Figure 27-12: Which two advancements made by annelids are shown in this picture?

How many different organ systems are represented?

Figure 27-13: What arthropod characteristics can you actually see in each of these photographs?

Figure 27-14: What features does the grasshopper have that humans also have?

What features do humans have that grasshoppers don't?

What features do grasshoppers have that humans don't?

Figure 27-15: What visible features justify placing this organism in the insect class?

Figure 27-16: What echinoderm features are visible in this sea star diagram?

How many organ systems are represented?

Is there any evidence of cephalization or segmentation?

Figure 27-17: What kind of symmetry is exhibited by this generalized chordate?

What evidence is there of cephalization?

What evidence is there of segmentation?

Is there a coelom?

Figure 27-18: If the adult tunicate lacks most chordate characteristics, why are tunicates considered chordates?

Figure 27-19: Compare this diagram with the generalized chordate in Figure 27-17. Why do you suppose they are so similar?

Figure 27-20: What part of the lamprey is shown in this figure?

Based on this photograph, why aren't lampreys considered radially symmetrical?

Figure 27-21: What vertebrate characteristics are visible?

What visible characteristics indicate these are bony fish (class Osteichthyes)?

What other bony fish features can not be seen?

Figure 27-22: What vertebrate characteristics are visible?

What visible characteristics indicate these are amphibians (class Amphibia)?

What other amphibian features can not be seen?

Figure 27-23: What vertebrate characteristics are visible?

What visible characteristics indicate these are reptiles (class Reptilia)?

What other reptilian features can not be seen?

Figure 27-24: What vertebrate characteristics are visible?

What visible characteristics indicate it is a bird (class Aves)?

What other avian features can not be seen?

What flight adaptations do birds have?

Figure 27-25: What vertebrate characteristics are visible?

What visible characteristics indicate these are mammals (class Mammalia)?

What other mammalian features can not be seen?

How are the three mammalian groups distinguished?

## Matching Questions

Write the letter of the phrase that best matches the numbered term on the left. Use each only once.

____ 1. radial symmetry
____ 2. bilateral symmetry
____ 3. cephalization
____ 4. body segment
____ 5. spongocoel
____ 6. polyp
____ 7. acoelomate
____ 8. pseudocoelomate
____ 9. exoskeleton
____10. radula
____11. parapodia
____12. metamorphosis
____13. tube feet
____14. swim bladder
____15. amphibian
____16. amniotic egg
____17. feather
____18. monotreme
____19. marsupial
____20. primate

a. unjointed appendages of some annelids
b. a cnidarian body form
c. having a body cavity not lined with peritoneum
d. a body plan that can be divided through a central point into many mirror images
e. egg-laying mammal
f. part of the echinoderm water vascular system
g. organ used for buoyancy in bony fish
h. one of repeating subunits in body construction
i. pouched mammal
j. a body plan that can be divided only into right and left halves
k. first vertebrate class to be terrestrial
l. an evolutionary development credited to reptiles
m. nonliving covering of arthropods
n. mammalian order that includes apes and monkeys
o. cavity characteristic of sponges
p. functions in insulation
q. developmental process involving a dramatic change in body form
r. lacking an internal body cavity
s. a structure containing chitinous teeth
t. evolutionary trend towards developing sense organs and nervous tissue at the front of an animal

## Multiple Choice Questions

1. Which IS NOT typical of animals with radial symmetry?
   a. They are capable of moving quickly through their environment.
   b. Sensory receptors are evenly distributed over body surfaces.
   c. Their body can be divided in more than one way to produce mirror images.
   d. Their body is usually cylindrical.
   e. They experience the environment equally from all directions.

2. Which IS NOT typical of animals with bilateral symmetry?
   a. They demonstrate cephalization.
   b. Their body can be divided in only one plane to produce mirror images.
   c. Their body has a definite front end to it.
   d. They are sessile.
   e. Sensory receptors are concentrated in one part of the body.

3. Which describes porifera?
   a. They have bilateral symmetry.
   b. They are predators.
   c. Choanocytes are specialized feeding cells.
   d. They have a tissue level of organization.
   e. They have a chitinous exoskeleton.

4. Which describes porifera?
   a. They have radial symmetry.
   b. They are filter feeders.
   c. Porocytes are cells specialized for motility.
   d. They have an organ level of organization.
   e. They have a calcium carbonate shell.

5. Which describes cnidarians?
   a. They have bilateral symmetry.
   b. They are herbivorous.
   c. All cnidarians go through a polyp stage at some point in the life cycle.
   d. The gastrovascular cavity is lined with mesoglea.
   e. They have a tissue level of organization.

6. Which describes cnidarians?
   a. They have some type of radial symmetry.
   b. They are autotrophic.
   c. All cnidarians go through a medusa stage at some point in the life cycle.
   d. Cnidocytes are found only in the gastrovascular cavity.
   e. They have an organ level of organization.

7. Which describes platyhelminths?
   a. They are pseudocoelomates.
   b. They have no symmetry.
   c. They have a brain and nerve cords.
   d. They have a tissue level of organization.
   e. All are phagocytic.

8. Which describes platyhelminths?
   a. They have a true coelom.
   b. They have radial symmetry.
   c. They have no nervous tissue.
   d. They have an organ level of organization.
   e. Some are free-living and others are parasitic.

9. Which describes nematodes?
   a. They are acoelomates.
   b. They have radial symmetry.
   c. They have a cartilagenous skeleton.
   d. They are unsegmented.
   e. They have jointed appendages.

10. Which describes nematodes?
   a. They are pseudocoelomates.
   b. They have no symmetry.
   c. They have spicules for skeletal support.
   d. They are segmented.
   e. They have unjointed appendages.

11. Which DOES NOT describe mollusks?
   a. Their internal organs are concentrated in a dorsal visceral mass.
   b. They belong to the second largest animal phylum.
   c. The mantle cavity is the site of gas exchange.
   d. Cephalopods are the most complex mollusks.
   e. They are segmented.

12. Which DOES NOT describe mollusks?
   a. Many have a radula with chitinous teeth.
   b. The mantle secretes a shell, if present.
   c. Most have a muscular foot for locomotion.
   d. They have jointed appendages.
   e. They are unsegmented.

13. Which describes annelids?
   a. They are segmented.
   b. They are acoelomates.
   c. They have jointed appendages.
   d. They have a chitinous exoskeleton.
   e. Adults are sessile.

14. Which describes annelids?
   a. They are unsegmented.
   b. They are pseudocoelomates.
   c. They have no appendages.
   d. Their body cavity acts as a hydrostatic skeleton.
   e. They evolved from a poriferan ancestor.

15. Which DOES NOT describe arthropods?
   a. The cuticle is nonliving.
   b. Lungs are found in terrestrial arthropods.
   c. They have a system level of organization.
   d. They have bilateral symmetry.
   e. Molting of the exoskeleton allows growth.

16. Which DOES NOT describe arthropods?
   a. They exhibit the greatest amount of diversity of any animal phylum.
   b. They are capable of quick movements.
   c. They are unsegmented.
   d. The cuticle acts as an exoskeleton.
   e. They have jointed appendages.

17. Which is a characteristic unique to chordates?
    a. bilateral symmetry
    b. true coelom
    c. notochord
    d. motility
    e. cephalization

18. Which is a characteristic unique to chordates?
    a. a tail
    b. segmentation
    c. a digestive tube with mouth and anus
    d. pharyngeal gill slits
    e. nervous tissue organized into a brain and nerve cords

19. Which group is characterized by three body segments, six jointed legs, wings, and an exoskeleton?
    a. cephalopods
    b. polychaetes
    c. insects
    d. crustaceans
    e. mollusks

20. Which group is characterized by a true coelom, segmentation, and unjointed appendages?
    a. polychaetes
    b. cephalopods
    c. nematodes
    d. chelicerates
    e. myriapods

21. Which group is characterized by no coelom, radial symmetry, cnidocytes, and a gastrovascular cavity?
    a. cnidarians
    b. gastropods
    c. cephalopods
    d. echinoderms
    e. crustaceans

22. Which group is characterized by bilateral symmetry and no coelom?
    a. cnidarians
    b. sponges
    c. nematodes
    d. flatworms
    e. mollusks

23. Which vertebrate class is characterized by gills, swim bladder and a bony skeleton?
    a. chondrichthyes
    b. osteichthyes
    c. amphibia
    d. reptilia
    e. aves

24. Which vertebrate class is terrestrial but must return to an aquatic environment to reproduce?
    a. chondrichthyes
    b. mammalia
    c. amphibia
    d. reptilia
    e. aves

25. Which vertebrate class has a single jaw bone, fur and nurses their young?
    a. mammalia
    b. osteichthyes
    c. amphibia
    d. reptilia
    e. aves

26. Which vertebrate class is characterized by a dry, scaly skin, and the production of an amniotic egg?
    a. chondrichthyes
    b. osteichthyes
    c. amphibia
    d. reptilia
    e. mammalia

27. Which vertebrate class is characterized by a cartilagenous skeleton in adults?
    a. chondrichthyes
    b. osteichthyes
    c. amphibia
    d. reptilia
    e. aves

28. Which vertebrate class produces an amniotic egg and maintains a constant body temperature?
    a. chondrichthyes
    b. osteichthyes
    c. amphibia
    d. mammalia
    e. aves

## Concept Map Construction

Construct a concept map for each group of terms. Be sure to include appropriate connector phrases. You may add other terms as necessary and use terms in the singular or plural form.

1. radial symmetry, segmentation, invertebrate, coelom, spongocoel
2. water vascular system, appendage, insect, radula, exoskeleton
3. amniotic egg, polyp, chordate, echinoderm, cnidaria

# Chapter 28

# Networks of Life: Ecology and the Biosphere

## Go Figure!

Figure 28-2d: Which line represents the equator?...the tropic of Cancer?...the tropic of Capricorn?

The three parallel lines represent the north to south range of tropical latitudes. What physical characteristic determines this range?

Are tropical rain forests restricted to this region?

Which biomes are not generally found in this region?

Figure 28-3: How do coastal autotrophs differ from pelagic autotrophs?

Figure 28-4: Why might these fish have eyes?

Figure 28-6: Describe potential environmental temperature differences between organisms living in the splash zone and those in the low intertidal zone.

Two filter feeding species live in the intertidal habitat, one in the high intertidal the other in the middle intertidal zone. How would this difference affect food availability?

Figure 28-8: What does the red line represent?

How does this line change from spring to summer to autumn?

What does the darkest green shaded layer under "Summer" represent?

What do the arrows represent under "Spring and autumn?"

Why is warmer water more buoyant?

What forces drive spring overturn?...autumn overturn?

Figure 28-10: Is it possible to have tundra at the equator according to this figure?

What will happen to the elevation of tree line as you move north from the equator?

What will happen to the elevation at which coniferous forest occurs as you move north from the equator?

Figures 28-11 through 28-19: Examine these biome photographs. Why do you suppose biomes are characterized by the predominant vegetation rather than by the animals?

## Matching Questions

Write the letter of the phrase that best matches the numbered term on the left. Use each only once.

____ 1. CFC
____ 2. ecosystem
____ 3. climate
____ 4. trade winds
____ 5. westerlies
____ 6. oligotrophic
____ 7. littoral zone
____ 8. epipelagic zone
____ 9. benthic zone
____10. plankton
____11. upwelling
____12. estuary
____13. middle intertidal zone
____14. limnetic zone
____15. shrubland
____16. coniferous forest
____17. overstory
____18. alpine tundra
____19. aestivation
____20. permafrost

a. defined as...includes living and non-living components of the environment
b. formed by air moving from 30° latitude toward 60° latitude
c. dominated by evergreen trees
d. covered by water 50% of the time
e. permanently frozen soil
f. region of shallow water at lake's edge
g. refers to sea floor
h. formed by air moving from 30° latitude toward the equator
i. found at high elevation above tree line
j. chlorofluorocarbon
k. upper layer of a tropical rain forest
l. surfacing of nutrient-ladened water
m. photic zone
n. a lake's open, lighted waters
o. behavioral adaptation of some desert animals
p. biome of short, woody plants
q. young, nutrient-poor lake
r. organisms that drift with ocean currents
s. prevailing weather in an area
t. region of mixed fresh water and salt water

## Multiple Choice Questions

1. Ecology includes the study of
   a. natural selection.
   b. adaptation.
   c. how many organisms occur in a given place.
   d. the origin of variability within a species.
   e. an organism's ancestral history.

2. Ecology includes the study of
   a. development.
   b. why organisms occur where they do.
   c. patterns of inheritance.
   d. origin of homologous structures.
   e. origin of island chains.

3. The earth's climates are primarily affected by the circulation patterns of
   a. the lithosphere.
   b. the benthic zone.
   c. the atmosphere.
   d. the bathypelagic zone.
   e. phytoplankton.

4. The earth's climates are primarily affected by the circulation patterns of
   a. acid rain.
   b. estuaries.
   c. the epipelagic zone.
   d. the oceans.
   e. the profundal zone.

5. During its summer, the Northern Hemisphere
   a. loses less heat than the Southern Hemisphere.
   b. is tipped towards the sun.
   c. is tipped away from the sun.
   d. gains heat from the Southern Hemisphere.
   e. is closer to the moon.

6. During its winter, the Northern Hemisphere
   a. loses less heat than the Southern Hemisphere.
   b. is tipped towards the sun.
   c. is tipped away from the sun.
   d. gains heat from the Southern Hemisphere.
   e. is closer to the moon.

Use the following information for questions 7 and 8.

Rising air creates areas of low atmospheric pressure. Areas of sinking air have high atmospheric pressure.

7. Which latitudes have prevailing high pressure?
   a. $0^{o}$ and $30^{o}$ north
   b. $30^{o}$ north and $30^{o}$ south
   c. $30^{o}$ south and $60^{o}$ north
   d. $60^{o}$ north and $60^{o}$ south
   e. $60^{o}$ north and $90^{o}$ north

8. Which latitudes have prevailing low pressure?
   a. $0^{\circ}$ and $30^{\circ}$ north
   b. $30^{\circ}$ north and $30^{\circ}$ south
   c. $30^{\circ}$ south and $60^{\circ}$ north
   d. $60^{\circ}$ north and $60^{\circ}$ south
   e. $60^{\circ}$ north and $90^{\circ}$ north

9. Coastal waters are generally very rich in life. Which is a factor important in allowing such species diversity?
   a. Coastal areas are generally very warm.
   b. All coastal waters surround coral reefs, and most reef species spend at least part of their life cycle there.
   c. They generally contain abundant nutrients from the neighboring land.
   d. They are generally very calm, non-turbulent areas.
   e. Hydrothermal vents provide coastal waters with an abundance of nutrients.

10. Coastal waters are generally very rich in life. Which is a factor important in allowing such species diversity?
   a. Coastal areas are generally very cold.
   b. All coastal waters surround estuaries, and most estuarine species spend at least part of their life cycle there.
   c. They generally are nutrient poor areas.
   d. Turbulence stirs the water and distributes nutrients to photosynthetic organisms.
   e. Coral reefs provide coastal waters with an abundance of nutrients.

11. Heterotrophs are found in the ____________ zone.
   a. epipelagic
   b. mesopelagic
   c. bathypelagic
   d. epipelagic and mesopelagic
   e. epipelagic, mesopelagic and bathypelagic

12. In which part of the pelagic zone is photosynthesis activity the greatest?
   a. epipelagic
   b. mesopelagic
   c. bathypelagic
   d. zooplankton
   e. phytoplankton

13. Which is an accurate comparison of a lake's limnetic and profundal zones?
   a. The limnetic zone is the area around the lake's edge, and the profundal zone is all other lighted areas.
   b. The limnetic zone is the area of the lake's open, lighted water, and the profundal zone is all other lighted areas.
   c. The limnetic zone is the area of the lake's open, lighted water, and the profundal zone is the area below the depth light penetrates.
   d. The limnetic zone is the area around the lake's edge, and the profundal zone is the area below the depth light penetrates.
   e. The profundal zone is the area around the lake's edge, and the limnetic zone is the area below which light penetrates.

14. Which is an accurate comparison of a lake's limnetic and profundal zones?
    a. Phytoplankton are the major photosynthetic producers of the limnetic zone, and rooted plants are the major photosynthetic producers of the profundal zone.
    b. Phytoplankton are the major photosynthetic producers of the limnetic zone, and there are no photosynthetic producers in the profundal zone.
    c. Rooted plants are the major photosynthetic producers of the limnetic zone, and phytoplankton are the major photosynthetic producers of the profundal zone.
    d. Floating multicellular algae are the major photosynthetic producers of the limnetic zone, and there are no photosynthetic producers in the profundal zone.
    e. Floating multicellular algae are the major photosynthetic producers of the profundal zone, and there are no photosynthetic producers in the limnetic zone.

15. Which biome supports the greatest species diversity?
    a. tropical rain forest
    b. deciduous forest
    c. coniferous forest
    d. grassland
    e. tundra

16. Which biome supports the least species diversity?
    a. tropical rain forest
    b. deciduous forest
    c. coniferous forest
    d. grassland
    e. tundra

17. In what way are tropical rain forests and coniferous forests similar?
    a. They have a comparable rate of decomposition.
    b. Both typically have three stories of plants.
    c. Both have great species diversity.
    d. Both have relatively nutrient poor soil.
    e. Over half of all species on earth are evenly divided between these two biomes.

18. In what way are tropical rain forests and coniferous forests similar?
    a. They have a comparable rate of leaf litter accumulation.
    b. Both typically have two stories of plants.
    c. Both have little species diversity.
    d. Both have relatively nutrient rich soil.
    e. Over one third of all species on earth are evenly divided between these two biomes.

19. Which biome is found in regions with a Mediterranean climate and is characterized by woody, perennial plants, shorter than trees with small, leathery leaves?
    a. tundra
    b. coniferous forest
    c. grassland
    d. shrubland
    e. desert

20. Which biome comprises 30 percent of the earth's terrestrial surface and occurs in areas with descending air?
    a. tundra
    b. coniferous forest
    c. grassland
    d. shrubland
    e. desert

21. Permafrost is characteristic of which biome?
    a. tropical rain forest
    b. deciduous forest
    c. coniferous forest
    d. grassland
    e. tundra

22. Which biome is characterized by plants that lose their leaves each autumn?
    a. tropical rain forest
    b. deciduous forest
    c. coniferous forest
    d. grassland
    e. tundra

## Concept Map Construction

Construct a concept map for each group of terms. Be sure to include appropriate connector phrases. You may add other terms as necessary and use terms in the plural or singular form.

1. osmoregulation, intertidal zone, low intertidal zone, estuary, filter feeder
2. photosynthesis, limnetic zone, mesotrophic, species diversity, salinity
3. ecosystem, biome, intertidal zone, stabilizing selection, tropical rain forest

# Chapter 29

# Natural Order of Life: Ecosystems and Communities

## Go Figure!

Figure 29-2: What would happen to developing frog eggs if environmental temperature went below 0°C?

What would happen to developing frog eggs if environmental temperature was constant at 5°C?

How would the rate of frog egg development compare at 8.75°C and 27°C?

What do tolerance ranges suggest about the environment of trout compared to that of frogs?

Figure 29-3: Turn your attention to the plot of two factors. If this organism is in a habitat with a temperature of 17°C and a yearly total rainfall of 62 millimeters, will it survive?

Will it survive in a habitat with a temperature of 32°C and a yearly total rainfall of 51 millimeters?

If a plot of temperature and yearly rainfall produces a point within the yellow area, will the organism live or die?

If a plot of temperature and yearly rainfall produces a point within the white area will the organism live or die?

Turn your attention to the plot of three factors. If a plot of temperature and yearly rainfall produces a point within the yellow area, will the organism live or die?

If a plot of temperature and yearly rainfall produces a point outside the yellow area, will the organism live or die?

Figure 29-4: Habitat destruction is causing massive species extinction. Which organisms should be able to better handle habitat destruction: those with a broad niche or those with a narrow niche?

Figure 29-5: Which trophic level contains autotrophic organisms?

Which trophic levels contain heterotrophic organisms?

What is the source of energy used in the first trophic level?

What is the source energy used in the second trophic level?

Figure 29-6: Five primary consumers are listed for manzanita. In what ways do their niches differ?

In what ways are they similar?

Since these consumers all feed on manzanita, are they in competition?

Figure 29-7: When the snake eats frogs, what consumer level does it represent?

When it eats field mice, what consumer level does it represent?

What level of consumer is the field mouse?

A field mouse will also eat grasshoppers. When it does, what consumer level does it represent?

Figure 29-8: What is meant by the phrase "DDT enters the food chain"?

Would the effects of DDT contamination be as severe if bioconcentration did not occur?

Figure 29-10: Which organisms add carbon dioxide to the atmosphere?

Which remove it?

How does carbon dioxide form bicarbonate in the lake?

Since combustion of fossil fuels releases $CO_2$ into the atmosphere that should have been released by decomposers anyway, why is it considered to be a problem?

Figure 29-11: What general type of organism plays a role in Cycle A <u>and</u> Cycle B?

How does the element nitrogen get passed from grass to cow?

Photosynthesis and respiration are antagonistic processes in the carbon cycle. What processes are antagonistic in the nitrogen cycle?

Figure 29-14: What do pioneer species in primary and secondary succession have in common?

## Matching Questions

Write the letter of the phrase that best matches the numbered term on the left. Use each only once.

____ 1. ecosystem

____ 2. abiotic

____ 3. primary producer

____ 4. detritus

____ 5. Victor Shelford

____ 6. Justus von Liebig

____ 7. detritus food web

____ 8. habitat

____ 9. fundamental niche

____10. realized niche

____11. biomass

____12. primary consumer

____13. pyramid of energy

____14. biogeochemical cycles

____15. nitrogen fixation

____16. nitrifying bacteria

____17. denitrification

____18. primary succession

____19. pioneer community

____20. secondary succession

a. generally credited with first stating the Law of the Minimum

b. small disintegrated pieces of organic wastes and dead organisms

c. term which applies specifically to the physical location of an organism

d. illustrates the rate at which energy moves through trophic levels

e. convert ammonia into nitrites

f. a progression of distinct communities in a previously disturbed habitat

g. first community in a new habitat

h. refers to all inanimate parts of an ecosystem

i. a progression of distinct communities in a new habitat

j. an organism which feeds upon producers

k. includes the community and the physical environment

l. elements passing back and forth between abiotic and biotic ecosystem components

m. an organism's potential niche

n. an organism's actual niche

o. generally credited with first stating the Theory of Tolerance

p. release of molecular nitrogen from nitrogen compounds

q. describes feeding relationships of decomposers

r. conversion of atmospheric nitrogen to ammonia

s. one criterion used in constructing ecological pyramids

t. autotroph

## Multiple Choice Questions

1. The primary ecological levels of organization include
   a. populations.
   b. individual organisms.
   c. organs.
   d. cells.
   e. molecules.

2. The primary ecological levels of organization include
   a. organ systems.
   b. tissues.
   c. subatomic particles.
   d. ecosystems.
   e. organelles.

3. The relationship between lake and cave ecosystems described in your text illustrates
   a. that ecosystems may function as separate units in terms of their energy use.
   b. that ecosystems may function as separate units in terms of their nutrient use.
   c. that ecosystems may depend on other ecosystems for exchange of materials and energy.
   d. that ecosystems may depend on other ecosystems for exchange of materials only.
   e. that the entire earth is one large ecosystem.

4. The description of a lake and the surrounding terrestrial ecosystem in your text illustrates
   a. that ecosystems may function as separate units in terms of their energy use.
   b. that ecosystems may function as separate units in terms of their nutrient use.
   c. that ecosystems may depend on other ecosystems for exchange of materials and energy.
   d. that ecosystems may depend on other ecosystems for exchange of materials only.
   e. that the entire earth is one large ecosystem.

5. Which is an example of an abiotic component affecting a biotic one?
   a. photosynthesis releasing molecular oxygen into the atmosphere
   b. detritus being added to soil
   c. a coral reef modifying ocean current direction
   d. soil phosphorus content limiting photosynthetic rate
   e. plant roots reducing the rate of soil erosion

6. Which is an example of a biotic component affecting an abiotic one?
   a. molecular oxygen content of soil affecting the abundance of soil nematodes
   b. a forest modifying the local humidity range
   c. ocean currents distribute larval forms along a coast
   d. soil iron content limiting the rate of photosynthetic rate
   e. soil erosion reducing soil fertility

7. The primary ecological role played by algae is
   a. primary producer.
   b. primary consumer.
   c. secondary consumer.
   d. decomposer.
   e. detritovore.

8. The primary ecological role played by fungi is
   a. primary producer.
   b. primary consumer.
   c. secondary consumer.
   d. tertiary consumer.
   e. decomposer.

9. Trout eggs will not develop if the temperature rises above 12°C. This illustrates
   a. the Theory of Tolerance.
   b. the Law of the Minimum.
   c. the full range of niche breadth.
   d. abiotic resources.
   e. biological magnification.

10. When nitrogen concentrations fall below a certain level, phytoplankton's photosynthetic rate is directly related to nitrogen concentration in the water. This illustrates
   a. the Theory of Tolerance.
   b. the Law of the Minimum.
   c. the full range of niche breadth.
   d. abiotic resources.
   e. biological magnification.

11. The dry, hot summers of Southern California result in a coastal chaparral biome made up of woody, drought resistant shrubs. This illustrates
   a. the Theory of Tolerance.
   b. the Law of the Minimum.
   c. the full range of niche breadth.
   d. the existence of ecological equivalents.
   e. biological magnification.

12. Coral reefs will not grow in water which falls below 16-18°C. This illustrates
   a. the Theory of Tolerance.
   b. the Law of the Minimum.
   c. the full range of niche breadth.
   d. the existence of ecological equivalents.
   e. biological magnification.

Use the following information for questions 13 and 14.

Barnacle larvae of the genus *Chthamalus* settle and metamorphose to adults over a wide intertidal range. When barnacles of the species *Balanus* are absent from the environment, *Chthamalus* adults survive within the entire intertidal range. When *Balanus* is present, *Chthamalus* adults are found only in the high intertidal zone.

13. Which statement is consistent with these observations?
   a. *Chthamalus* reduces *Balanus*' hypervolume.
   b. The presence of *Balanus* reduces the realized niche of *Chthamalus*.
   c. *Chthamalus* and *Balanus* have identical potential niches.
   d. *Chthamalus* reduces the tolerance range of *Balanus*.
   e. *Balanus* reduces the tolerance range of *Chthamalus*.

14. Which statement is consistent with these observations?
   a. *Chthamalus* and *Balanus* show niche overlap.
   b. The presence of *Chthamalus* reduces the realized niche of *Balanus*.
   c. *Chthamalus* and *Balanus* form a food chain.
   d. This situation demonstrates ecological succession.
   e. Communities containing *Chthamalus* and *Balanus* are climax communities.

15. Which moves through a food chain?
   a. nutrients
   b. primary producers
   c. guilds
   d. hypervolumes
   e. succession

16. Which moves through a food chain?
    a. primary consumers
    b. detritovores
    c. energy
    d. sunlight
    e. geochemical cycles

17. Under which conditions would you expect to see an ecological pyramid of numbers with a smaller number of producers than primary consumers?
    a. in an ecosystem with small producers and large primary consumers
    b. in an ecosystem with small producers and small primary consumers
    c. in an ecosystem with large producers and large primary consumers
    d. in an ecosystem with large producers and small primary consumers
    e. in an ecosystem with grasses being eaten by mice

18. Under which conditions would an ecological pyramid of biomass have a smaller biomass of producers than primary consumers?
    a. *producers*–large, bottom dwelling algae with no seasonal change in abundance
       *primary consumers*–small and seasonal in abundance
    b. *producers*–large, bottom dwelling algae with no seasonal change in abundance
       *primary consumers-large* and seasonal in abundance
    c. *producers*–small, bottom dwelling algae with no seasonal change in abundance
       *primary consumers*–small and seasonal in abundance
    d. *producers*–small and have significant changes in seasonal abundance
       *primary consumers*–large with no seasonal changes in abundance except for migrations
    e. *producers*–small and have large changes in seasonal abundance
       *primary consumers*–large with seasonal changes in abundance excluding migrations

19. How do the carbon and nitrogen biogeochemical cycles compare?
    a. Both use photosynthesis to move atoms between biotic and abiotic components of an ecosystem.
    b. Both directly use aerobic respiration to move atoms between biotic and abiotic components of an ecosystem.
    c. Both have an atmospheric reservoir of element.
    d. Both are dependent upon bacteria and cyanobacteria to move the element between biotic and abiotic components of the cycle.
    e. One has a gaseous phase to the cycle, the other does not.

20. How do the phosphorus and nitrogen biogeochemical cycles compare?
    a. Both use photosynthesis to move atoms between biotic and abiotic components of an ecosystem.
    b. Both use respiration to move atoms between biotic and abiotic components of an ecosystem.
    c. Both have a large atmospheric reservoir of the element.
    d. Both are dependent upon bacteria and cyanobacteria to move the element between biotic and abiotic components of the cycle.
    e. One has a gaseous phase to the cycle, the other does not.

21. How do climax and pioneer communities compare?
    a. Both communities remain constant in species composition for long periods of time.
    b. Both communities have a large net accumulation of biomass.
    c. The plant species in the climax community are generally longer lived than in the pioneer community.
    d. Pioneer communities tend to have more complex food webs than climax communities.
    e. Producers in both communities use a large amount of their energy intake for reproduction.

22. How do climax and pioneer communities compare?
    a. The climax community remains constant in species composition, the pioneer community does not.
    b. Only the climax community has a large net accumulation of biomass.
    c. The plant species in the pioneer community are generally longer lived than in the climax community
    d. Pioneer communities tend to have more primary consumers than climax communities.
    e. Only the climax community autotrophs use a large proportion of their energy intake for reproduction.

## Concept Map Construction

Construct a concept map for each group of terms. Be sure to include appropriate connector phrases. You may add other terms as necessary and use terms in the plural or singular form.

1. ecosystem, niche, resource, temperature, food chain
2. photosynthesis, hydrologic cycle, phosphorus cycle, detritovore, food web
3. biome, pioneer community, species diversity, biomass, tolerance range

# Chapter 30

# Life's Interactions: Communities

## Go Figure!

Figure 30-2: Compare the diagram on the left with the one on the right. Competition between species A and B is greater on the right than the left. In what other ways do these two graphs differ?

Could competition between these two species increase without the other differences?

Figure 30-3: Describe how the food resource of the Bay-breasted warbler differs from that of the Yellow-rumped warbler.

Describe how food resources of Blackburnian and Bay-breasted warblers differ. What type of resource partitioning is this?

Figure 30-4: On the average, how many years after the hare population peaked did the lynx population peak?

On the average, how many years after the hare population bottomed out did the lynx population reach their minimum?

What might account for the fact that between 1900 and 1906 hares reached their lowest level for the entire sample period, but lynx reached their third highest peak?

What was the greatest number of hare pelts taken during this period?

What was the smallest number of hare pelts taken during this period?

Figures 30-6 through 30-10: Suggest a sequence of events (consistent with what is currently known about biology) that would lead to the development of each style of protection.

Figure 30-11: Suggest a sequence of events (consistent with what is currently known about biology) that would lead to the development of this example of Batesian mimicry.

How does Batesian mimicry differ from Müllerian mimicry?

Figure 30-12: Since the pearl fish has poor eyesight, suggest other ways in which it might locate its host sea cucumber.

Figure 30-15: What type of symbiotic relationship exists between termites and their gut protozoa?

## Matching Questions

Write the letter of the phrase that best matches the numbered term on the left. Use each only once.

_____ 1. herbivore

_____ 2. exploitative competition

_____ 3. niche overlap

_____ 4. interference competition

_____ 5. territoriality

_____ 6. resource partitioning

_____ 7. prey

_____ 8. carnivore

_____ 9. omnivore

_____10. escape response

_____11. aposematic coloring

_____12. allelochemical

_____13. disruptive coloration

_____14. Batesian mimicry

_____15. Müllerian mimicry

_____16. parasite

_____17. social parasitism

_____18. epiphyte

_____19. host

_____20. commensalism

a. plant eater

b. a form of indirect competition

c. dividing an environmental resource between two species

d. organism that eats plants or animals

e. compound which discourages predators

f. disguises the shape of an organism

g. feeds on a host

h. results in interspecific competition

i. behavior that reduces competition within a species

j. an interaction in which one organism profits and the other is unaffected

k. specific term which applies to a palatable mimic resembling a distasteful model

l. organism that eats animals

m. specific term which applies to two distasteful organisms resembling each other

n. exploiting the behavior of the host

o. acts as a warning

p. loss of a lizard's tail is an example

q. organism that suffers the effects of a parasite

r. an example of a commensal organism

s. directly blocking access of another species to environmental resources is an example

t. an organism fed upon by another organism

## Multiple Choice Questions

1. Which is an example of symbiosis?
   a. competition
   b. predation
   c. commensalism
   d. herbivory
   e. competitive exclusion

2. Which IS NOT an example of symbiosis?
   a. parasitism
   b. mimicry
   c. protocooperation
   d. endosymbiosis
   e. mutualism

3. The greater the niche overlap of two species
   a. the more likely they are to enter into a symbiotic relationship.
   b. the more likely they are to show competition between each other.
   c. the more likely they are to show competition within their own species.
   d. the less likely they are to show competition between each other.
   e. the less likely they are to show competition within their own species.

4. Which is likely to be the most intense competitive situation?
   a. competition within a species (intraspecific competition)
   b. competition between species (interspecific competition)
   c. commensalistic
   d. exploitative
   e. allelopathy

5. Your text describes deep-rooted tamarisk trees tapping groundwater supplies in the California desert, thereby reducing water supplies to native mesquite and desert willows. This is an example of
   a. intraspecific competition.
   b. exploitative competition.
   c. interference competition.
   d. territoriality.
   e. character displacement.

6. Your text describes hyenas driving vultures away from the remains of a zebra. This is an example of
   a. intraspecific competition.
   b. exploitative competition.
   c. interference competition.
   d. territoriality.
   e. character displacement.

7. Domestic goats were introduced to the island of Abingdon in the Galapagos Archipelago in 1957. The goats reduced island food supplies suitable for native tortoises, and by 1962, the tortoises were no longer found on the island. This is an example of
   a. allelopathy.
   b. protocooperation.
   c. commensalism.
   d. competitive exclusion.
   e. resource partitioning.

8. Your text describes five species of North American warblers that feed on insects in slightly different parts of the same tree. This is an example of
   a. allelopathy.
   b. protocooperation.
   c. commensalism.
   d. competitive exclusion.
   e. resource partitioning.

9. Two species of nuthatches with partial overlapping niches also are found in many of the same geographical ranges. Where their ranges overlap, bill size and coloration are strikingly different. Where ranges do not overlap, bills are remarkable similar. This is an example of
   a. intraspecific competition.
   b. exploitative competition.
   c. interference competition.
   d. territoriality.
   e. character displacement.

10. When two hypothetical species occupy the same area, they show character displacement, apparently in response to competition. When the two species are examined in parts of their ranges where they do not overlap, what are morphological data about this character likely to reveal?
    a. The two species show even greater differences than where they are cohabitants.
    b. The two species show differences similar to those shown where they are cohabitants.
    c. The two species show fewer differences than where they are cohabitants.
    d. They show exploitative competition.
    e. They show cryptic coloration.

11. Your text classifies "playing dead" as
    a. a group escape adaptation.
    b. an individual escape adaptation.
    c. a defense adaptation.
    d. a camouflage escape adaptation.
    e. Müllerian mimicry.

12. Your text classifies aposematic coloration as
    a. a group escape adaptation.
    b. an individual escape adaptation.
    c. a defense adaptation.
    d. a camouflage escape adaptation.
    e. Müllerian mimicry.

13. When a smelt fish detects an approaching predator, it releases warning chemicals into the water. This is
    a. a group escape adaptation.
    b. an individual escape adaptation.
    c. a defense adaptation.
    d. a camouflage escape adaptation.
    e. Müllerian mimicry.

14. A noxious beetle has a bright orange wingcase with black tips. These are very similar in appearance to the wingcases of another noxious beetle. This is an example of
    a. a group escape adaptation.
    b. an individual escape adaptation.
    c. Batesian mimicry.
    d. a camouflage escape adaptation.
    e. Müllerian mimicry.

15. A mimic gains selective advantage by resembling a distasteful or poisonous species. What is the likely consequence if the mimic is neither distasteful nor poisonous, and its population size has been reduced to one-tenth its normal size?
    a. The selective advantage gained by mimicry will be increased.
    b. The selective advantage gained by mimicry will be decreased.
    c. The mimic will experience an increase in predation.
    d. The distasteful model will experience a decrease in predation.
    e. The selective advantage gained by mimicry will not change.

16. A mimic gains selective advantage by resembling a distasteful or poisonous species. What is the likely consequence if the mimic is neither distasteful nor poisonous, and is ten times as abundant as normal?
    a. The selective advantage gained by mimicry will be increased.
    b. The selective advantage gained by mimicry will be decreased.
    c. The mimic will experience a decrease in predation.
    d. The distasteful model will experience an increase in predation.
    e. The selective advantage gained by mimicry will not change.

17. The remora/shark relationship is cited by your text as demonstrating
    a. commensalism.
    b. parasitism.
    c. mutualism.
    d. protocooperation.
    e. allelopathy.

18. A lichen is cited by your text as demonstrating
    a. commensalism.
    b. parasitism.
    c. mutualism.
    d. protocooperation.
    e. allelopathy.

19. How does commensalism differ from protocooperation?
    a. In commensalism, one organism benefits and the other is harmed, while in protocooperation both are harmed.
    b. In commensalism, one organism benefits and the other is harmed, while in protocooperation both benefit.
    c. In commensalism, both organisms benefit, while in protocooperation both are harmed.
    d. In commensalism, one organism benefits and the other is unaffected, while in protocooperation both benefit.
    e. In commensalism, both organisms benefit, while in protocooperation one organism benefits.

20. How do mutualism and protocooperation compare?
    a. In mutualism one organism benefits and the other is harmed, while in protocooperation both are harmed.
    b. In mutualism one organism benefits and the other is harmed, while in protocooperation both benefit.
    c. In mutualism both organisms benefit, while in protocooperation both are harmed.
    d. In mutualism one organism benefits and the other is unaffected, while in protocooperation both benefit.
    e. In both mutualism and protocooperation, both organisms benefit.

## Concept Map Construction

Construct a concept map for each group of terms. Be sure to include appropriate connector phrases. You may add other terms as necessary and use terms in the plural or singular form.

1. tolerance range, competition, predation, symbiosis, cryptic coloration
2. coevolution, lichen, exploitative competition, prey, protocooperation
3. limiting factor, character displacement, Batesian mimicry, predator, parasite

# Chapter 31

# Animal Behavior

## Go Figure!

Figure 31-1: Describe how you would attempt to demonstrate that the courtship of a stickleback is an innate rather than a learned behavior.

Of what selective advantage are FAPs?

Figure 31-4: What type of learning does this represent if the monkeys learned the behavior from other monkeys?

What type of learning does this represent if the monkeys learned the behavior from the caretaker?

Figure 31-7: Suggest controlled experiments that could have been used to determine the meaning of the bee's dance.

Figure 31-8: How might you attempt to collect evidence to support the statement "grooming originally functioned only for skin care..."?

How might grooming behavior confer selective advantage upon the organism doing the grooming?

What human behaviors might be modified grooming behaviors?

What purpose do these human behaviors currently serve?

## Matching Questions

Write the letter of the phrase that best matches the numbered term on the left. Use each only once.

____ 1. ethology
____ 2. innate behavior
____ 3. sign stimulus
____ 4. releaser
____ 5. chain of reactions
____ 6. egg-laying hormone
____ 7. habituation
____ 8. conditioned reflex
____ 9. insight learning
____10. operant conditioning
____11. filial imprinting
____12. critical period
____13. recruitment
____14. pheromone
____15. altruism
____16. alarm call

a. general term for a single, genetically controlled, highly stereotyped, species specific animal response

b. chemical which increases the rate of action potentials generated from *Aplysia*'s abdominal ganglion

c. a learned response resulting in a new association between a stimulus and a response

d. the limited developmental period during which filial imprinting may occur

e. specific example of altruism

f. a sudden solution to a problem without trial and error procedures

g. a sign stimulus from a member of the same species

h. general term for something in the environment which elicits a fixed action pattern

i. performance of a service which benefits a conspecific at some cost to the one who performs the service

j. the study of animal behavior

k. the simplest form of learning

l. term which applies when individuals of a species are brought together to perform a specific task

m. requires that a behavior originally occur spontaneously

n. in young birds, a process of developing a preference for following their mother

o. an intricate sequence of fixed action patterns

p. a chemical signal which influences the behavior of others of the same species

## Multiple Choice Questions

1. In general, behaviors that are precisely specified by genes are
   a. often those which must be expressed in nearly perfect form on the first trial.
   b. often those which appear in intricate patterns called a chain of reactions.
   c. associated with egg laying or mating.
   d. easily habituated.
   e. called ELHs.

2. Innate behaviors are
   a. often those which must be expressed in nearly perfect form on the first trial.
   b. often those which appear in intricate patterns called a chain of reactions.
   c. associated with egg laying or mating.
   d. easily habituated.
   e. called ELHs.

3. A fixed action pattern is
   a. a specific example of habituation.
   b. beneficial in that it eliminates a response to frequently occurring stimuli that have no bearing on an animal's welfare.
   c. beneficial in that it focuses attention and energy use on important aspects of the environment.
   d. a motor response triggered by a sign stimulus.
   e. a form of learning.

4. Classical conditioning is
   a. a specific example of habituation.
   b. beneficial in that it eliminates a response to frequently occurring stimuli that have no bearing on an animal's welfare.
   c. beneficial in that it focuses attention and energy use on important aspects of the environment.
   d. a motor response triggered by a sign stimulus.
   e. a form of learning.

5. Your text cites the retrieval of an egg by a female greylag goose as evidence that
   a. habituation occurs.
   b. behaviors are initiated by sign stimuli.
   c. once a FAP begins, it continues to completion.
   d. learning is an important part of all behaviors.
   e. unconditioned stimuli initiate operant conditioning.

6. Your text cites the retrieval of an egg by a female greylag goose as evidence that
   a. habituation does not occurs.
   b. sign stimuli are initiated by behaviors.
   c. once an FAP begins, the sequence of actions is virtually identical.
   d. learning is an important part of all innate behaviors.
   e. conditioned stimuli initiate operant conditioning.

7. When Niko Tinbergen and co-workers presented dummy male sticklebacks to real males, the researchers were able to demonstrate that
   a. sticklebacks defend their territories vigorously.
   b. a red tint on the undersurface of a model stickleback habituates real males to the presence of other real males.
   c. a red tint on the undersurface acts as a territory defense behavior releaser.
   d. the size and shape of the model acts as a territory defense behavior releaser.
   e. territory defense behavior is an example of a conditioned stimulus.

8. When Niko Tinbergen and co-workers presented dummy male sticklebacks to real males, the researchers were able to demonstrate that
   a. sticklebacks defend their territories vigorously when females are present.
   b. a red tint on the undersurface of a model stickleback conditions real males to the presence of other real males.
   c. a red tint on the undersurface does not acts as a territory defense behavior releaser.
   d. the size and shape of the model does not act as a territory defense behavior releaser.
   e. territory defense behavior is an example of an unconditioned stimulus.

9. It has been shown that in stickleback courtship
   a. a female entering a male's territory and exposing her egg-swollen abdomen acts as a releaser initiating male gamete production.
   b. a female entering a male's territory and exposing her egg-swollen abdomen acts as a releaser initiating a zig-zag dance by the male.
   c. a zig-zag dance by a female initiates approach behavior by a male.
   d. a zig-zag dance by a male initiates approach behavior by another male.
   e. a female and male are never in the nest at the same time.

10. It has been shown that in stickleback courtship
    a. "head-up" posture acts as a releaser initiating dorsal spine raising by a female.
    b. "head-up" posture acts as a releaser initiating dorsal spine raising by a male.
    c. a zig-zag dance by a male initiates dorsal spine raising behavior by another male.
    d. a zig-zag dance by a male initiates approach behavior by a female.
    e. a female and male are always in the nest at the same time.

11. Which is NOT an example of learning?
    a. A clamworm repeatedly encounters a shadow passing overhead. At first, it quickly retreats into its burrow when the shadow appears. Eventually the worm no longer responds to the shadow.
    b. Many times food was presented to Pavlov's dogs following the ringing of a bell. Eventually the dogs would salivate when they heard the bell.
    c. A male robin entering the territory of another is vigorously attacked. A tuft of red feathers placed in a male robin's territory is also vigorously attacked.
    d. A hungry rat is placed in a box with a lever. If it presses the lever, food enters the cage. At first, the rat touches the lever while randomly exploring the box. Given sufficient time, the rat begins pressing the lever in a nonrandom manner.
    e. A gosling follows the first moving object it sees after hatching.

12. Which is NOT an example of learning?
    a. habituation
    b. fixed action pattern
    c. classical conditioning
    d. operant conditioning
    e. imprinting

13. Captive chimpanzees who had no previous direct or indirect experience stacking boxes to retrieve food stacked boxes in order to reach a banana. This is an example of
    a. habituation.
    b. insight learning.
    c. classical conditioning.
    d. operant conditioning.
    e. social learning.

14. Identical twins were separated at birth. Both are now eight years old. Joey has been with their mother since he was released from the hospital and Rufus has been with their father. Neither youngster has spent any time with the other parent. Joey has a pet dog and thinks cats are for wimps; Rufus has a pet cat and is afraid of dogs. If their pet preferences were learned, it is most likely a consequence of
    a. habituation.
    b. insight learning.
    c. classical conditioning.
    d. operant conditioning.
    e. social learning.

15. One hypothesis concerning this behavior states that sensory and motor stimulation of this activity may cause the formation of cerebellar synaptic networks. What is being described?
    a. habituation
    b. play
    c. filial imprinting
    d. territorial behavior
    e. fixed action patterns

16. This type of learning is distinguished from others by the speed with which it occurs, the limited time during which it can occur, and that it occurs without any obvious reward.
    a. habituation
    b. play
    c. filial imprinting
    d. territorial behavior
    e. fixed action patterns

17. Which is a benefit of group living?
    a. increased exposure to parasites
    b. decreased transmission of disease
    c. increased energy used by group raising of young
    d. increased protection from predators
    e. decreased mating efficiency

18. Which is a cost of group living?
    a. decreased exposure to parasites
    b. increased transmission of disease
    c. decreased energy used by individuals in raising young
    d. decreased hunting efficiency by predators
    e. decreased mating efficiency

19. According to your text, which mode of communication is best over long distances in an open terrestrial environment?
    a. visual
    b. sound
    c. pheromones
    d. body contact
    e. scent trails

20. According to your text, which mode of communication is best over long distances in the open ocean?
    a. visual
    b. sound
    c. pheromones
    d. body contact
    e. scent trails

## Concept Map Construction

Construct a concept map for each group of terms. Be sure to include appropriate connector phrases. You may add other terms as necessary and use terms in the plural or singular form.

1. innate behavior, chain of reactions, learning, habituation, unconditioned stimulus
2. behavior, FAP, gene, transcription, ELH
3. natural selection, altruism, kin selection, cooperative mate selection, learning

# Chapter 32

# Population Dynamics

## Go Figure!

Figure 32-2: Please look at the Type II survivorship curve. When a population has gone from birth to 20 percent of its life span, what fraction has survived?

What is the magnitude of change when a Type II population has gone from 20 to 50 percent of its life span?

Please look at the Type III survivorship curve. When a population has gone from birth to 20 percent of its life span, what fraction has survived?

What is the magnitude of change when a Type III population has gone from 20 to 50 percent of its life span?

Figure 32-4: This growth curve is for a population which doubles each time its members synchronously reproduce. How would the curve be affected if population members synchronously had 6 offspring?

This population growth curve is for an organism capable of reproducing every 20 minutes. How would the curve be affected if it was for an organism that had offspring every 6 weeks?

Figure 32-5: What conditions must exist for dieback #1 to occur?

What conditions might account for dieback #3?

Based upon the graph in Figure 32-5b, did this *Daphnia* population go extinct?

Figure 32-6: What factors may account for the sheep population increasing above average carrying capacity?

What might account for decreases below the carrying capacity average?

What impact would you guess the introduction of sheep had on native Tasmanian herbivores?

Figure 32-7: Was there any human population growth between 100,000 and 40,000 years ago?

Did the human population grow between 100 and 10 years ago?

Figure 32-8: What is happening to the length of time it takes to add one billion people to the earth?

What limiting factors kept the human population small early in its history? How have those limiting factors changed?

What has happened to the earth's human carrying capacity between 10,000 BC and the present?

Figure 32-9: How do these two age-sex structure diagrams differ for prereproductive age groups?

Are there any similarities?

How do they differ for reproductive age groups?

Are there any similarities?

How do they differ for postreproductive age groups?

Are there any similarities?

Suppose developing countries were able to convince their citizens to have only two children (enough only to replace themselves). Why would the population continue to grow for several decades before finally stabilizing?

## Matching Questions

Write the letter of the phrase that best matches the numbered term on the left. Use each only once.

____ 1. population density
____ 2. population growth rate
____ 3. fertility rate
____ 4. carrying capacity
____ 5. clumped distribution in a population
____ 6. random distribution in a population
____ 7. natality
____ 8. zero population growth
____ 9. Type I survivorship curve
____10. biotic potential
____11. sigmoid growth curve
____12. emigration
____13. environmental resistance
____14. J-shaped growth curve
____15. dieback
____16. density-dependent limiting factor
____17. percent annual increase
____18. doubling time

a. maximum population size an environment can support indefinitely

b. birth rate

c. a population's innate capacity to increase

d. number of individuals of a single species per unit area

e. distribution which may result when water is available at a limited number of sites

f. a mortality induced decrease in population size

g. length of time required for population size to increase by a factor of two

h. change in population density per unit of time

i. number of organisms added to a population per 100 individuals in the population

j. sum of all limiting factors

k. results when population mortality is low for most of the species expected life span

l. eventually exhibited by a population occupying a new environment

m. individuals moving out of an area

n. a distribution which may result when water is available uniformly in an environment

o. disease

p. produced by a population in exponential growth

q. average number of children born to each reproductive age woman

r. occurs when a population's gains and losses are equal

## Multiple Choice Questions

1. All living organisms within an ecosystem are called
   a. a biome.
   b. a biosphere.
   c. a community.
   d. abiotic conditions.
   e. a population.

2. All of the yellow pine trees in a yellow pine forest comprise
   a. a biome.
   b. a biosphere.
   c. a community.
   d. abiotic conditions.
   e. a population.

3. How do population density and growth rate compare?
   a. Population density is expressed as organisms per unit area and growth rate as clumped or random.
   b. Population density is expressed as change per unit time and growth rate as number per unit area.
   c. Population density is measured as organisms per unit area and growth rate as change per unit time.
   d. Population density is determined directly by natality and growth rate by emigration.
   e. Population density is determined directly by emigration and growth rate by natality.

4. How do population density and distribution compare?
   a. Population density is expressed as organisms per unit area and distribution as clumped or random.
   b. Population density is expressed as change per unit time and distribution as number per unit area.
   c. Population density is measured as organisms per unit area and distribution as change per unit time.
   d. Population density is determined directly by natality and distribution by emigration.
   e. Population density is determined directly by emigration and distribution by natality.

5. A plant population may be spread through the environment in clumps. Which factor probably DOES NOT contribute to this?
   a. Favorable conditions for germination are arranged in patches.
   b. Favorable conditions for photosynthesis are arranged in patches.
   c. Some plant seeds are released in groups.
   d. Some plants reproduce asexually by means of runners.
   e. Some plants release chemicals from their roots which inhibit growth of other members of their species.

6. Some organisms are uniformly spread through the environment. Which factor probably DOES NOT contribute to this?
   a. environmental conditions are highly variable throughout the area
   b. territoriality
   c. release of chemicals from the roots of some plants inhibit growth of other members of their species
   d. the presence of chemicals in a tree's leaf and bark inhibits growth of other members of their species
   e. environmental conditions are more or less the same throughout the habitat

Use the following information for questions 7 and 8.

The sum of natality and immigration minus the sum of mortality and emigration gives a population's net growth. Natality = N, Immigration = I, Mortality = M, and Emigration = E. Each is expressed as a percent of the starting population size of 100 rabbits.

7. Which values would lead to growth of the rabbit population?
   a. N=20, I=0, M=4, E=16
   b. N=11, I=1, M=3, E=12
   c. N=2, I=15, M=12, E=0
   d. N=6, I=2, M=7, E=2
   e. N=6, I=8, M=15, E=4

8. Which values would lead to zero population growth of the rabbit population?
   a. N=20, I=0, M=4, E=16
   b. N=11, I=1, M=3, E=12
   c. N=2, I=15, M=12, E=0
   d. N=6, I=2, M=7, E=2
   e. N=6, I=8, M=15, E=4

9. How do Type I and Type II survivorship curves compare?
   a. Type I shows high mortality and Type II shows low mortality in the early years of life
   b. Both show low mortality in the early years of life.
   c. Both show high mortality in the early years of life.
   d. Type I shows increasing mortality with age and Type II shows a constant mortality.
   e. Type I shows increasing mortality with age and Type II shows decreasing mortality with age.

10. How do Type I and Type III survivorship curves compare?
   a. Type I shows high mortality and Type III shows low mortality in the early years of life
   b. Both show low mortality in the early years of life.
   c. Both show high mortality in the early years of life.
   d. Type I shows increasing mortality with age and Type III shows a constant mortality.
   e. Type I shows increasing mortality with age and Type III shows decreasing mortality with age.

Use the following information for questions 11 and 12.

R represents a population's net reproductive rate. If population size is constant, R = 1. When it is greater than 1, the population is increasing in size, and when it is less than 1, the population is decreasing. Use the formula $N_1 = N_0 R$ to calculate the value of R for the growth of *E. coli* plotted in Figure 32-4 of your text. ($N_1$ equals population size after 1 interval of time [in this case 20 minutes] and $N_0$ equals the initial population size).

11. Calculate R based upon an increase from 32 ($N_0$) to 64 ($N_1$).
   a. 1
   b. 2
   c. 3
   d. 4
   e. 5

12. Calculate R based upon an increase from 2048 ($N_0$) to 4096 ($N_1$).
   a. 1
   b. 2
   c. 3
   d. 4
   e. 5

Use the following information for questions 13 and 14.

The following are age-sex structures for different populations. Assume no migration, or that immigration and emigration are equal and do not alter the age-sex structure.

a.

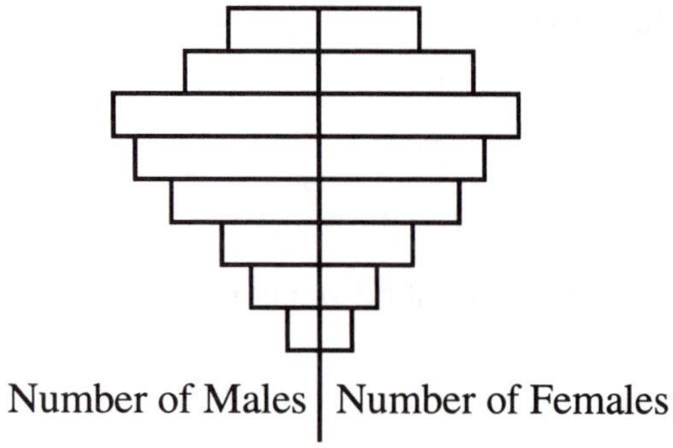


b.

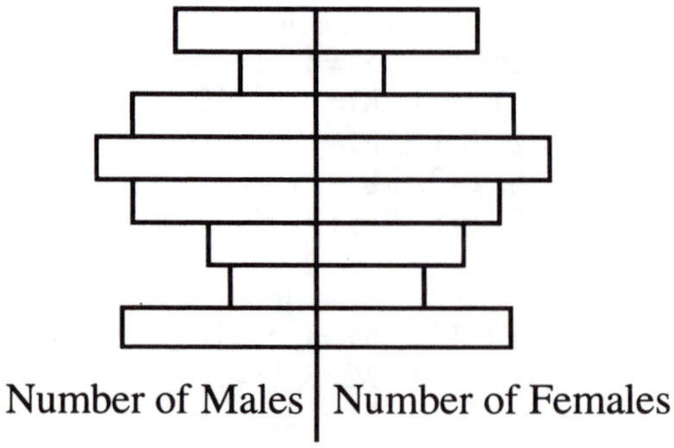


c.

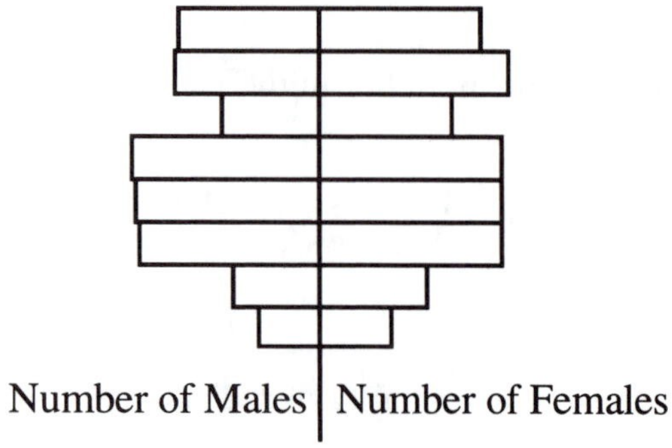


d.

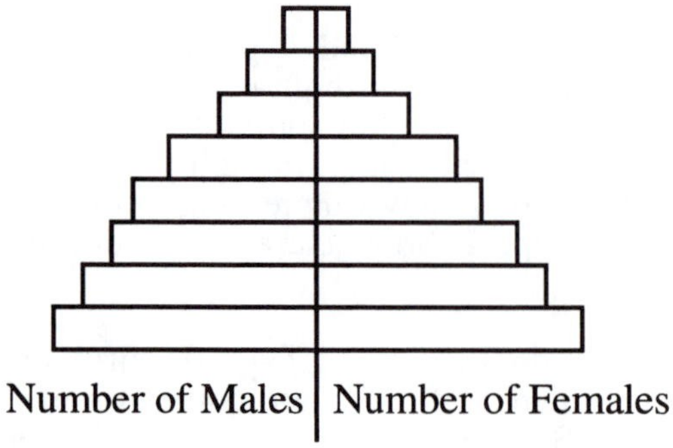


e.

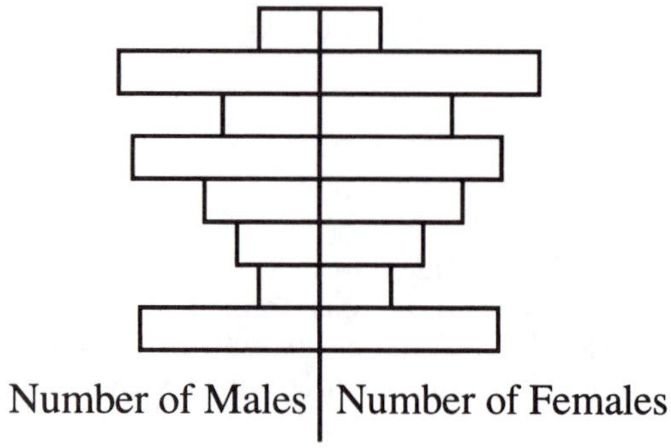


13. Which population will grow the fastest?

14. Which population will experience a sustained decrease in size?

15. How are environmental resistance and carrying capacity related?
    a. Carrying capacity determines the amount of environmental resistance.
    b. Environmental resistance determines an environment's carrying capacity.
    c. Environmental resistance describes the abiotic portion of an environment's carrying capacity.
    d. Carrying capacity describes the biotic portion of environmental resistance.
    e. They are different terms for the same thing.

16. How do J-shaped and S-shaped population growth compare?
    a. J-shaped growth levels off at carrying capacity and S-shaped growth does not.
    b. Both begin with a slow rate of population increase and enter a period of exponential growth.
    c. J-shaped growth begins slowly and S-shaped begins rapidly.
    d. J-shaped growth enters an exponential growth phase and S-shaped does not.
    e. J-shaped growth levels off after an initial exponential phase and S-shaped does not.
17. Which is an example of a density-dependent limiting factor?
    a. disease
    b. fire
    c. volcanic lava flow
    d. mud slide
    e. application of insecticide to a crop

18. Which is an example of a density-independent limiting factor?
    a. competition
    b. predation
    c. infanticide
    d. flood
    e. stress

19. Please examine Table 32-1 What is Uganda's doubling time?
    a. 3.0 years
    b. 3.3 years
    c. 30 years
    d. 30.3 years
    e. 33.3 years

20. Please examine Table 32-1. What is the United Kingdom's doubling time?
    a. 5 years
    b. 50 years
    c. 55 years
    d. 500 years
    e. 550 years

## Concept Map Construction

Construct a concept map for each group of terms. Be sure to include appropriate connector phrases. You may add other terms as necessary and use terms in the plural or singular form.

1. limiting factor, density-dependent factor, zero population growth, natality, age-sex structure
2. carrying capacity, doubling time, tropical rain forest, resource depletion, fertility rate
3. niche, environmental resistance, exponential growth, Law of the Minimum, ecological pyramid

# Appendix

## Answers to Selected Study Guide Questions

### Chapter 1

#### Matching Questions

1. e
3. i
5. g
7. d
9. q
11. m
13. j
15. t
17. o
19. k

#### Multiple Choice Questions

1. c
3. d
5. c
7. c
9. c
11. a
13. a
15. b
17. b
19. b
20. d
21. a
22. b
23. d
24. b

### Chapter 2

#### Matching Questions

1. o
3. j
5. e
7. u
9. h
11. m
13. t
15. f
17. q
19. a
21. c

#### Multiple Choice Questions

1. a
3. b
5. c
7. d
9. b
11. b
13. c
15. d
17. a
19. a
21. c
23. a
25. b
27. d
29. c
31. a
33. d
35. a

### Chapter 3

#### Matching Questions

1. m
3. b
5. l
7. e
9. d
11. h
13. u
15. g
17. o
19 a
21. f

#### Multiple Choice Questions

1. c
3. c
5. b
7. e
9. d
11. e
13. d
15. d
17. a
19. a
21. c
23. d
25. e
27. c

## Chapter 4

### Matching Questions

1. a
3. d
5. b
7. q
9. p
11. o
13. n
15. c
17. g

### Multiple Choice Questions

1. d
3. d
5. b
7. c
9. d
11. d
13. a
15. a
17. a
19. b

## Chapter 5

### Matching Questions

1. f
3. a
5. c
7. b
9. p
11. q
13. o
15. k
17. n

### Multiple Choice Questions

1. c
3. b
5. a
7. d
9. d
11. d
13. e
15. d
17. a
19. e
21. a
23. b

## Chapter 6

### Matching Questions

1. g
3. b
5. p
7. q
9. h
11. l
13. o
15. k
17. i

### Multiple Choice Questions

1. a
3. c
5. a
7. e
9. c
11. b
13. b
15. a
17. b
19. c
21. c
23. e

## Chapter 7

### Matching Questions

1. b
3. a
5. c
7. d
9. j
11. n
13. m
15. o
17. p
19. s

### Multiple Choice Questions

1. b
3. d
5. d
7. e
9. b
11. c
13. a
15. b
17. c
19. c
21. a
23. c
25. c
27. d
29. a
31. e

## Chapter 8

### Matching Questions

1. d
3. a
5. h
7. g
9. i
11. k
13. p
15. q
17. s
19. j

### Multiple Choice Questions

1. a
3. d
5. a
7. b
9. a
10. c
11. c
12. d
17. c
19. c
21. a
23. c
25. c
27. c
29. b
31. c
32. d
33. c
37. b
39. b
41. b

## Chapter 9

### Matching Questions

1. h
3. t
5. k
7. j
9. c
11. e
13. d
15. b
17. m
19. g

### Multiple Choice Questions

1. b
3. d
5. a
7. e
9. c
11. b
13. e
15. b
17. a
19. c
20. c
21. c
22. e

## Chapter 10

### Matching Questions

1. e
3. p
5. r
7. n
9. c
11. o
13. h
15. q
17. d

### Multiple Choice Questions

1. d
3. c
5. a
7. d
9. b
11. d
13. b
15. a
17. b
19. c

## Chapter 11

### Matching Questions

1. e
3. a
5. b
7. c
9. s
11. i
13. j
15. q
17. l
19. p

### Multiple Choice Questions

1. a
2. c
3. c
7. b
8. c
9. a
10. e
15. d
17. d
19. b
21. c
23. b
25. c
27. b
29. c

## Chapter 12

### Matching Questions

1. a
3. g
5. l
7. d
9. e
11. i
13. p
15. s
17. t
19. u
21. v
23. q

### Multiple Choice Questions

1. d
3. b
5. d
7. b
9. c
11. e
13. d
15. c
17. b
19. d
21. a
23. d
25. d
27. c
29. b
31. d
33. b
35. c
37. c

## Chapter 13

### Matching Questions

1. i
3. e
5. m
7. f
9. b
11. v
13. k
15. c
17. h
19. u
21. l
23. p

### Multiple Choice Questions

1. c
3. b
5. c
7. d
9. d
11. d
13. b
15. c
17. a
19. b
21. a
23. b
25. a
27. c
28. d

## Chapter 14

### Matching Questions

1. j
3 e
5. a
7. c
9. m
11. f
13. g
15. o
17. q

### Multiple Choice Questions

1. b
3. c
5. a
7. a
9. c
11. c
13. b
15. c
17. b
19. c
21. c
23. c

## Chapter 15

### Matching Questions

1. g
3. a
5. e
7. v
9. c
11. l
13. s
15. k
17. j
19. u
21. d
23. o

### Multiple Choice Questions

1. a
3. d
5. c
7. d
9. b
11. e
13. b
15. e
17. c
19. d
21. a
23. b
25. a
27. d
29. c
31. b
33. b
35. e

## Chapter 16

### Matching Questions

1. l
3. r
5. b
7. t
9. e
11. d
13. p
15. g
17. m
19. o

### Multiple Choice Questions

1. c
3. d
5. d
7. c
9. a
11. e
13. c
15. a
17. e
19. a
21. b

## Chapter 17

### Matching Questions

1. g
3. i
5. n
7. b
9. c
11. d
13. e
15. j
17. p
19. f

### Multiple Choice Questions

1. c
3. a
5. e
7. b
9. a
11. b
13. d
15. c
17. b
19. b

## Chapter 18

### Matching Questions

1. c
3. f
5. a
7. n
9. m
11. b
13. e
15. d
17. r
19. k

### Multiple Choice Questions

1. d
3. d
5. d
7. c
9. b
11. b
13. d
15. e
17. c
19. b
21. a

## Chapter 19

### Matching Questions

1. n
3. c
5. e
7. q
9. d
11. g
13. u
15. t
17. h
19. k
21. p

### Multiple Choice Questions

1. d
3. b
5. b
7. a
9. b
11. e
13. e
15. c
17. a
19. c
21. a
23. c
25. d
27. a
29. a
31. b
33. a

## Chapter 20

### Matching Questions

1. a
3. t
5. f
7. j
9. r
11. d
13. m
15. q
17. c
19. b

### Multiple Choice Questions

1. a
3. e
5. d
7. d
9. b
11. e
13. c
15. b
17. d
19. c

## Chapter 21

### Matching Questions

1. d
3. p
5. j
7. k
9. q
11. l
13. f
15. n
17. g
19. a
21. s
23. u

### Multiple Choice Questions

1. c
3. b
5. a
7. a
9. a
11. a
13. d
15. c
17. a
19. b
21. a
23. c
25. d
27. c
29. e
31. d
33. a

## Chapter 22

### Matching Questions

1. i
3. b
5. d
7. n
9. q
11. h
13. m
15. e
17. k
19. c

### Multiple Choice Questions

1. a
3. d
4. a
7. a
9. a
11. d
13. b
15. c
17. a
19. b
21. b
23. a

## Chapter 23

### Matching Questions

1. s
3. a
5. r
7. g
9. b
11. d
13. p
15. o
17. q
19. f
21. i

### Multiple Choice Questions

1. d
3. e
5. b
7. a
9. c
11. c
13. b
15. c
17. e
19. e

## Chapter 24

### Matching Questions

1. i
3. j
5. c
7. o
9. a
11. f
13. h
15. p
17. g

### Multiple Choice Questions

1. d
3. d
5. a
7. e
9. a
11. d
13. b
15. d
17. a
19. b

## Chapter 25

### Matching Questions

1. g
3. d
5. j
7. n
9. k
11. p
13. f
15. r
17. b
19. t
21. i

### Multiple Choice Questions

1. e
3. d
5. d
7. c
9. b
11. a
13. b
15. d
17. d
19. b
21. e
23. e
25. b
27. c
29. d

## Chapter 26

### Matching Questions

1. j
3. a
5. q
7. b
9. o
11. t
13. k
15. f
17. n
19. g

### Multiple Choice Questions

1. c
3. b
5. e
7. d
9. c
11. c
13. b
15. a
17. b
19. c
21. b

## Chapter 27

### Matching Questions

1. d
3. t
5. o
7. r
9. m
11. a
13. f
15. k
17. p
19. i

### Multiple Choice Questions

1. a
3. c
5. e
7. c
9. d
11. e
13. a
15. b
17. c
19. c
21. a
23. b
25. a
27. a

## Chapter 28

### Matching Questions

| | |
|---|---|
| 1. | j |
| 3. | s |
| 5. | b |
| 7. | f |
| 9. | g |
| 11. | l |
| 13. | d |
| 15. | p |
| 17. | k |
| 19. | o |

### Multiple Choice Questions

| | |
|---|---|
| 1. | c |
| 3. | c |
| 5. | b |
| 7. | b |
| 9. | c |
| 11. | e |
| 13. | c |
| 15. | a |
| 17. | d |
| 19. | d |
| 21. | e |

## Chapter 29

### Matching Questions

| | |
|---|---|
| 1. | k |
| 3. | t |
| 5. | o |
| 7. | q |
| 9. | m |
| 11. | s |
| 13. | d |
| 15. | r |
| 17. | p |
| 19. | g |

### Multiple Choice Questions

| | |
|---|---|
| 1. | a |
| 3. | c |
| 5. | d |
| 7. | a |
| 9. | a |
| 11. | b |
| 13. | b |
| 15. | a |
| 17. | d |
| 19. | c |
| 21. | c |

## Chapter 30

### Matching Questions

| | |
|---|---|
| 1. | a |
| 3. | h |
| 5. | i |
| 7. | t |
| 9. | d |
| 11. | o |
| 13. | f |
| 15. | m |
| 17. | n |
| 19. | q |

### Multiple Choice Questions

| | |
|---|---|
| 1. | c |
| 3. | b |
| 5. | b |
| 7. | d |
| 9. | e |
| 11. | b |
| 13. | a |
| 15. | a |
| 17. | a |
| 19. | d |

## Chapter 31

### Matching Questions

1. j
3. h
5. o
7. k
9. f
11. n
13. l
15. i

### Multiple Choice Questions

1. a
3. d
5. c
7. c
9. b
11. c
13. b
15. b
17. d
19. c

## Chapter 32

### Matching Questions

1. d
3. q
5. e
7. b
9. k
11. l
13. j
15. f
17. i

### Multiple Choice Questions

1. c
3. c
5. e
7. c
9. d
11. b
13. d
15. b
17. a
19. e